Xingzhi Zhan

Matrix Inequalities

 Springer

Author

Xingzhi ZHAN
Institute of Mathematics
Peking University
Beijing 100871, China

E-mail: zhan@math.pku.edu.cn

Cataloging-in-Publication Data applied for

Die Deutsche Bibliothek - CIP-Einheitsaufnahme

Zhan, Xingzhi:
Matrix inequalities / Xingzhi Zhan. - Berlin ; Heidelberg ; New York ;
Barcelona ; Hong Kong ; London ; Milan ; Paris ; Tokyo : Springer, 2002
  (Lecture notes in mathematics ; Vol. 1790)
  ISBN 3-540-43798-3

Mathematics Subject Classification (2000):
15-02, 15A18, 15A60, 15A45, 15A15, 47A63

ISSN 0075-8434
ISBN 3-540-43798-3 Springer-Verlag Berlin Heidelberg New York

Springer-Verlag Berlin Heidelberg New York a member of BertelsmannSpringer
Science + Business Media GmbH

http://www.springer.de

© Springer-Verlag Berlin Heidelberg 2002
Printed in Germany

Typesetting: Camera-ready TeX output by the author

SPIN: 10882616   41/3142/du-543210 - Printed on acid-free paper

Lecture Notes in Mathematics

1790

Editors:
J.-M. Morel, Cachan
F. Takens, Groningen
B. Teissier, Paris

Springer
Berlin
Heidelberg
New York
Barcelona
Hong Kong
London
Milan
Paris
Tokyo

Preface

Matrix analysis is a research field of basic interest and has applications in scientific computing, control and systems theory, operations research, mathematical physics, statistics, economics and engineering disciplines. Sometimes it is also needed in other areas of pure mathematics.

A lot of theorems in matrix analysis appear in the form of inequalities. Given any complex-valued function defined on matrices, there are inequalities for it. We may say that matrix inequalities reflect the quantitative aspect of matrix analysis. Thus this book covers such topics as norms, singular values, eigenvalues, the permanent function, and the Löwner partial order.

The main purpose of this monograph is to report on recent developments in the field of matrix inequalities, with emphasis on useful techniques and ingenious ideas. Most of the results and new proofs presented here were obtained in the past eight years. Some results proved earlier are also collected as they are both important and interesting.

Among other results this book contains the affirmative solutions of eight conjectures. Many theorems unify previous inequalities; several are the culmination of work by many people. Besides frequent use of operator-theoretic methods, the reader will also see the power of classical analysis and algebraic arguments, as well as combinatorial considerations.

There are two very nice books on the subject published in the last decade. One is *Topics in Matrix Analysis* by R. A. Horn and C. R. Johnson, Cambridge University Press, 1991; the other is *Matrix Analysis* by R. Bhatia, GTM 169, Springer, 1997. Except a few preliminary results, there is no overlap between this book and the two mentioned above.

At the end of every section I give notes and references to indicate the history of the results and further readings.

This book should be a useful reference for research workers. The prerequisites are linear algebra, real and complex analysis, and some familiarity with Bhatia's and Horn-Johnson's books. It is self-contained in the sense that detailed proofs of all the main theorems and important technical lemmas are given. Thus the book can be read by graduate students and advanced undergraduates. I hope this book will provide them with one more opportunity to appreciate the elegance of mathematics and enjoy the fun of understanding certain phenomena.

I am grateful to Professors T. Ando, R. Bhatia, F. Hiai, R. A. Horn, E. Jiang, M. Wei and D. Zheng for many illuminating conversations and much help of various kinds.

This book was written while I was working at Tohoku University, which was supported by the Japan Society for the Promotion of Science. I thank JSPS for the support. I received warm hospitality at Tohoku University.

Special thanks go to Professor Fumio Hiai, with whom I worked in Japan. I have benefited greatly from his kindness and enthusiasm for mathematics.

I wish to express my gratitude to my son Sailun whose unique character is the source of my happiness.

Sendai, December 2001   *Xingzhi Zhan*

Contents

1. Inequalities in the Löwner Partial Order

Throughout we consider square complex matrices. Since rectangular matrices can be augmented to square ones with zero blocks, all the results on singular values and unitarily invariant norms hold as well for rectangular matrices. Denote by M_n the space of $n \times n$ complex matrices. A matrix $A \in M_n$ is often regarded as a linear operator on $\mathbb{C}^n$ endowed with the usual inner product $\langle x, y \rangle \equiv \sum_j x_j \bar{y}_j$ for $x = (x_j), y = (y_j) \in \mathbb{C}^n$. Then the conjugate transpose A^* is the adjoint of A. The Euclidean norm on $\mathbb{C}^n$ is $\|x\| = \langle x, x \rangle^{1/2}$. A matrix $A \in M_n$ is called *positive semidefinite* if

$$\langle Ax, x \rangle \geq 0 \quad \text{for all } x \in \mathbb{C}^n. \tag{1.1}$$

Thus for a positive semidefinite A, $\langle Ax, x \rangle = \langle x, Ax \rangle$. For any $A \in M_n$ and $x, y \in \mathbb{C}^n$, we have

$$4 \langle Ax, y \rangle = \sum_{k=0}^{3} i^k \langle A(x + i^k y), x + i^k y \rangle,$$

$$4 \langle x, Ay \rangle = \sum_{k=0}^{3} i^k \langle x + i^k y, A(x + i^k y) \rangle$$

where $i = \sqrt{-1}$. It is clear from these two identities that the condition (1.1) implies $A^* = A$. Therefore a positive semidefinite matrix is necessarily Hermitian.

In the sequel when we talk about matrices $A, B, C, \ldots$ without specifying their orders, we always mean that they are of the same order. For Hermitian matrices G, H we write $G \leq H$ or $H \geq G$ to mean that $H - G$ is positive semidefinite. In particular, $H \geq 0$ indicates that H is positive semidefinite. This is known as the *Löwner partial order*; it is induced in the real space of (complex) Hermitian matrices by the cone of positive semidefinite matrices. If H is positive definite, that is, positive semidefinite and invertible, we write $H > 0$.

Let $f(t)$ be a continuous real-valued function defined on a real interval Ω and H be a Hermitian matrix with eigenvalues in Ω. Let $H = U \operatorname{diag}(\lambda_1, \ldots, \lambda_n) U^*$ be a spectral decomposition with U unitary. Then the *functional calculus* for H is defined as

$$f(H) \equiv U \operatorname{diag}(f(\lambda_1), \dots, f(\lambda_n))U^*. \tag{1.2}$$

This is well-defined, that is, $f(H)$ does not depend on particular spectral decompositions of H. To see this, first note that (1.2) coincides with the usual polynomial calculus: If $f(t) = \sum_{j=0}^{k} c_j t^j$ then $f(H) = \sum_{j=0}^{k} c_j H^j$. Second, by the Weierstrass approximation theorem, every continuous function on a finite closed interval Ω is uniformly approximated by a sequence of polynomials. Here we need the notion of a norm on matrices to give a precise meaning of approximation by a sequence of matrices. We denote by $\|A\|_\infty$ the spectral (operator) norm of A: $\|A\|_\infty \equiv \max\{\|Ax\| : \|x\| = 1, x \in \mathbb{C}^n\}$. The spectral norm is submultiplicative: $\|AB\|_\infty \le \|A\|_\infty \|B\|_\infty$. The positive semidefinite square root $H^{1/2}$ of $H \ge 0$ plays an important role.

Some results in this chapter are the basis of inequalities for eigenvalues, singular values and norms developed in subsequent chapters. We always use capital letters for matrices and small letters for numbers unless otherwise stated.

1.1 The Löwner-Heinz inequality

Denote by I the identity matrix. A matrix C is called *a contraction* if $C^*C \le I$, or equivalently, $\|C\|_\infty \le 1$. Let $\rho(A)$ be the spectral radius of A. Then $\rho(A) \le \|A\|_\infty$. Since AB and BA have the same eigenvalues, $\rho(AB) = \rho(BA)$.

Theorem 1.1 (Löwner-Heinz) *If $A \ge B \ge 0$ and $0 \le r \le 1$ then*

$$A^r \ge B^r. \tag{1.3}$$

Proof. The standard continuity argument is that in many cases, e.g., the present situation, to prove some conclusion on positive semidefinite matrices it suffices to show it for positive definite matrices by considering $A + \epsilon I$, $\epsilon \downarrow 0$. Now we assume $A > 0$.

Let Δ be the set of those $r \in [0,1]$ such that (1.3) holds. Obviously $0, 1 \in \Delta$ and Δ is closed. Next we show that Δ is convex, from which follows $\Delta = [0,1]$ and the proof will be completed. Suppose $s, t \in \Delta$. Then

$$A^{-s/2} B^s A^{-s/2} \le I, \quad A^{-t/2} B^t A^{-t/2} \le I$$

or equivalently $\|B^{s/2} A^{-s/2}\|_\infty \le 1$, $\|B^{t/2} A^{-t/2}\|_\infty \le 1$. Therefore

$$
\begin{aligned}
\|A^{-(s+t)/4} B^{(s+t)/2} A^{-(s+t)/4}\|_\infty &= \rho(A^{-(s+t)/4} B^{(s+t)/2} A^{-(s+t)/4}) \\
&= \rho(A^{-s/2} B^{(s+t)/2} A^{-t/2}) \\
&= \|A^{-s/2} B^{(s+t)/2} A^{-t/2}\|_\infty \\
&= \|(B^{s/2} A^{-s/2})^* (B^{t/2} A^{-t/2})\|_\infty \\
&\le \|B^{s/2} A^{-s/2}\|_\infty \|B^{t/2} A^{-t/2}\|_\infty \\
&\le 1.
\end{aligned}
$$

Thus $A^{-(s+t)/4}B^{(s+t)/2}A^{-(s+t)/4} \leq I$ and consequently $B^{(s+t)/2} \leq A^{(s+t)/2}$, i.e., $(s+t)/2 \in \Delta$. This proves the convexity of Δ.   □

How about this theorem for $r > 1$? The answer is negative in general. The example

$$A = \begin{bmatrix} 2 & 1 \\ 1 & 1 \end{bmatrix}, \quad B = \begin{bmatrix} 1 & 0 \\ 0 & 0 \end{bmatrix}, \quad A^2 - B^2 = \begin{bmatrix} 4 & 3 \\ 3 & 2 \end{bmatrix}$$

shows that $A \geq B \geq 0 \not\Rightarrow A^2 \geq B^2$.

The next result gives a conceptual understanding, and this seems a typical way of mathematical thinking.

We will have another occasion in Section 4.6 to mention the notion of a C^*-algebra, but for our purpose it is just M_n. Let $\mathcal{A}$ be a Banach space over $\mathbb{C}$. If $\mathcal{A}$ is also an algebra in which the norm is submultiplicative: $\|AB\| \leq \|A\| \|B\|$, then $\mathcal{A}$ is called a *Banach algebra*. An *involution* on $\mathcal{A}$ is a map $A \mapsto A^*$ of $\mathcal{A}$ into itself such that for all $A, B \in \mathcal{A}$ and $\alpha \in \mathbb{C}$

(i) $(A^*)^* = A$;   (ii) $(AB)^* = B^* A^*$;   (iii) $(\alpha A + B)^* = \bar{\alpha} A^* + B^*$.

A C^*-*algebra* $\mathcal{A}$ is a Banach algebra with involution such that

$$\|A^* A\| = \|A\|^2 \quad \text{for all } A \in \mathcal{A}.$$

An element $A \in \mathcal{A}$ is called *positive* if $A = B^* B$ for some $B \in \mathcal{A}$.

It is clear that M_n with the spectral norm and with conjugate transpose being the involution is a C^*-algebra. Note that the Löwner-Heinz inequality also holds for elements in a C^*-algebra and the same proof works, since every fact used there remains true, for instance, $\rho(AB) = \rho(BA)$.

Every element $T \in \mathcal{A}$ can be written uniquely as $T = A + iB$ with A, B Hermitian. In fact $A = (T + T^*)/2$, $B = (T - T^*)/2i$. This is called the *Cartesian decomposition* of T.

We say that $\mathcal{A}$ is *commutative* if $AB = BA$ for all $A, B \in \mathcal{A}$.

Theorem 1.2 *Let $\mathcal{A}$ be a C^*-algebra and $r > 1$. If $A \geq B \geq 0$, $A, B \in \mathcal{A}$ implies $A^r \geq B^r$, then $\mathcal{A}$ is commutative.*

Proof. Since $r > 1$, there exists a positive integer k such that $r^k > 2$. Suppose $A \geq B \geq 0$. Use the assumption successively k times we get $A^{r^k} \geq B^{r^k}$. Then apply the Löwner-Heinz inequality with the power $2/r^k < 1$ to obtain $A^2 \geq B^2$. Therefore it suffices to prove the theorem for the case $r = 2$.

For any $A, B \geq 0$ and $\epsilon > 0$ we have $A + \epsilon B \geq A$. Hence by assumption, $(A + \epsilon B)^2 \geq A^2$. This yields $AB + BA + \epsilon B^2 \geq 0$ for any $\epsilon > 0$. Thus

$$AB + BA \geq 0 \quad \text{for all } A, B \geq 0. \tag{1.4}$$

Let $AB = G + iH$ with G, H Hermitian. Then (1.4) means $G \geq 0$. Applying this to $A, BAB,$

$$A(BAB) = G^2 - H^2 + i(GH + HG) \tag{1.5}$$

gives $G^2 \geq H^2$. So the set

$$\Gamma \equiv \{\alpha \geq 1 : G^2 \geq \alpha H^2 \text{ for all } A, B \geq 0 \text{ with } AB = G + iH\}$$

where $G + iH$ is the Cartesian decomposition, is nonempty. Suppose Γ is bounded. Then since Γ is closed, it has a largest element λ. By (1.4) $H^2(G^2 - \lambda H^2) + (G^2 - \lambda H^2)H^2 \geq 0$, i.e.,

$$G^2 H^2 + H^2 G^2 \geq 2\lambda H^4. \tag{1.6}$$

From (1.5) we have $(G^2 - H^2)^2 \geq \lambda(GH + HG)^2$, i.e.,

$$G^4 + H^4 - (G^2 H^2 + H^2 G^2)$$
$$\geq \lambda[GH^2 G + HG^2 H + G(HGH) + (HGH)G].$$

Combining this inequality, (1.6) and the inequalities $GH^2 G \geq 0$, $G(HGH) + (HGH)G \geq 0$ (by (1.4) and $G \geq 0$), $HG^2 H \geq \lambda H^4$ (by the definition of λ) we obtain

$$G^4 \geq (\lambda^2 + 2\lambda - 1)H^4.$$

Then applying the Löwner-Heinz inequality again we get

$$G^2 \geq (\lambda^2 + 2\lambda - 1)^{1/2} H^2$$

for all G, H in the Cartesian decomposition $AB = G + iH$ with $A, B \geq 0$. Hence $(\lambda^2 + 2\lambda - 1)^{1/2} \in \Gamma$, which yields $(\lambda^2 + 2\lambda - 1)^{1/2} \leq \lambda$ by definition. Consequently $\lambda \leq 1/2$. This contradicts the assumption that $\lambda \geq 1$. So Γ is unbounded and $G^2 \geq \alpha H^2$ for all $\alpha \geq 1$, which is possible only when $H = 0$. Consequently $AB = BA$ for all positive A, B. Finally by the Cartesian decomposition and the fact that every Hermitian element is a difference of two positive elements we conclude that $XY = YX$ for all $X, Y \in \mathcal{A}$. $\square$

Since M_n is noncommutative when $n \geq 2$, we know that for any $r > 1$ there exist $A \geq B \geq 0$ but $A^r \not\geq B^r$.

Notes and References. The proof of Theorem 1.1 here is given by G. K. Pedersen [79]. Theorem 1.2 is due to T. Ogasawara [77].

1.2 Maps on Matrix Spaces

A real-valued continuous function $f(t)$ defined on a real interval Ω is said to be *operator monotone* if

$$A \leq B \quad \text{implies} \quad f(A) \leq f(B)$$

for all such Hermitian matrices A, B of all orders whose eigenvalues are contained in Ω. f is called *operator convex* if for any $0 < \lambda < 1$,

$$f(\lambda A + (1 - \lambda)B) \leq \lambda f(A) + (1 - \lambda)f(B)$$

holds for all Hermitian matrices A, B of all orders with eigenvalues in Ω. f is called *operator concave* if $-f$ is operator convex.

Thus the Löwner-Heinz inequality says that the function $f(t) = t^r$, $(0 < r \leq 1)$ is operator monotone on $[0, \infty)$. Another example of operator monotone function is $\log t$ on $(0, \infty)$ while an example of operator convex function is $g(t) = t^r$ on $(0, \infty)$ for $-1 \leq r \leq 0$ or $1 \leq r \leq 2$ [17, p.147].

If we know the formula

$$t^r = \frac{\sin r\pi}{\pi} \int_0^\infty \frac{s^{r-1}t}{s+t} ds \quad (0 < r < 1)$$

then Theorem 1.1 becomes quite obvious. In general we have the following useful integral representations for operator monotone and operator convex functions. This is part of Löwner's deep theory [17, p.144 and 147] (see also [32]).

Theorem 1.3 *If f is an operator monotone function on $[0, \infty)$, then there exists a positive measure μ on $[0, \infty)$ such that*

$$f(t) = \alpha + \beta t + \int_0^\infty \frac{st}{s+t} d\mu(s) \tag{1.7}$$

where α is a real number and $\beta \geq 0$. If g is an operator convex function on $[0, \infty)$ then there exists a positive measure μ on $[0, \infty)$ such that

$$g(t) = \alpha + \beta t + \gamma t^2 + \int_0^\infty \frac{st^2}{s+t} d\mu(s) \tag{1.8}$$

where α, β are real numbers and $\gamma \geq 0$.

The three concepts of operator monotone, operator convex and operator concave functions are intimately related. For example, a nonnegative continuous function on $[0, \infty)$ is operator monotone if and only if it is operator concave [17, Theorem V.2.5].

A map $\Phi : M_m \to M_n$ is called *positive* if it maps positive semidefinite matrices to positive semidefinite matrices: $A \geq 0 \Rightarrow \Phi(A) \geq 0$. Denote by I_n the identity matrix in M_n. Φ is called *unital* if $\Phi(I_m) = I_n$.

We will first derive some inequalities involving unital positive linear maps, operator monotone functions and operator convex functions, then use these results to obtain inequalities for matrix Hadamard products.

The following fact is very useful.

Lemma 1.4 *Let $A > 0$. Then*

$$\begin{bmatrix} A & B \\ B^* & C \end{bmatrix} \geq 0$$

if and only if the Schur complement $C - B^*A^{-1}B \geq 0$.

Lemma 1.5 *Let Φ be a unital positive linear map from M_m to M_n. Then*

$$\Phi(A^2) \geq \Phi(A)^2 \quad (A \geq 0), \tag{1.9}$$

$$\Phi(A^{-1}) \geq \Phi(A)^{-1} \quad (A > 0). \tag{1.10}$$

Proof. Let $A = \sum_{j=1}^m \lambda_j E_j$ be the spectral decomposition of A, where $\lambda_j \geq 0 \, (j = 1, \ldots, m)$ are the eigenvalues and $E_j \, (j = 1, \ldots, m)$ are the corresponding eigenprojections of rank one with $\sum_{j=1}^m E_j = I_m$. Then since $A^2 = \sum_{j=1}^m \lambda_j^2 E_j$ and by unitality $I_n = \Phi(I_m) = \sum_{j=1}^m \Phi(E_j)$, we have

$$\begin{bmatrix} I_n & \Phi(A) \\ \Phi(A) & \Phi(A^2) \end{bmatrix} = \sum_{j=1}^m \begin{bmatrix} 1 & \lambda_j \\ \lambda_j & \lambda_j^2 \end{bmatrix} \otimes \Phi(E_j),$$

where $\otimes$ denotes the Kronecker (tensor) product. Since

$$\begin{bmatrix} 1 & \lambda_j \\ \lambda_j & \lambda_j^2 \end{bmatrix} \geq 0$$

and by positivity $\Phi(E_j) \geq 0 \, (j = 1, \ldots, m)$, we have

$$\begin{bmatrix} 1 & \lambda_j \\ \lambda_j & \lambda_j^2 \end{bmatrix} \otimes \Phi(E_j) \geq 0,$$

$j = 1, \ldots, m$. Consequently

$$\begin{bmatrix} I_n & \Phi(A) \\ \Phi(A) & \Phi(A^2) \end{bmatrix} \geq 0$$

which implies (1.9) by Lemma 1.4.

In a similar way, using

$$\begin{bmatrix} \lambda_j & 1 \\ 1 & \lambda_j^{-1} \end{bmatrix} \geq 0$$

we can conclude that

$$\begin{bmatrix} \Phi(A) & I_n \\ I_n & \Phi(A^{-1}) \end{bmatrix} \geq 0$$

which implies (1.10) again by Lemma 1.4. □

Theorem 1.6 *Let Φ be a unital positive linear map from M_m to M_n and f an operator monotone function on $[0, \infty)$. Then for every $A \geq 0$,*

$$f(\Phi(A)) \geq \Phi(f(A)).$$

Proof. By the integral representation (1.7) it suffices to prove

$$\Phi(A)[sI + \Phi(A)]^{-1} \geq \Phi[A(sI + A)^{-1}], \quad s > 0.$$

Since $A(sI + A)^{-1} = I - s(sI + A)^{-1}$ and similarly for the left side, this is equivalent to

$$[\Phi(sI + A)]^{-1} \leq \Phi[(sI + A)^{-1}]$$

which follows from (1.10). $\qquad\square$

Theorem 1.7 *Let Φ be a unital positive linear map from M_m to M_n and g an operator convex function on $[0, \infty)$. Then for every $A \geq 0$,*

$$g(\Phi(A)) \leq \Phi(g(A)).$$

Proof. By the integral representation (1.8) it suffices to show

$$\Phi(A)^2 \leq \Phi(A^2) \tag{1.11}$$

and

$$\Phi(A)^2[sI + \Phi(A)]^{-1} \leq \Phi[A^2(sI + A)^{-1}], \quad s > 0. \tag{1.12}$$

(1.11) is just (1.9). Since

$$A^2(sI + A)^{-1} = A - sI + s^2(sI + A)^{-1},$$

$$\Phi(A)^2[sI + \Phi(A)]^{-1} = \Phi(A) - sI + s^2[sI + \Phi(A)]^{-1},$$

(1.12) follows from (1.10). This completes the proof. $\qquad\square$

Since $f_1(t) = t^r$ $(0 < r \leq 1)$ and $f_2(t) = \log t$ are operator monotone functions on $[0, \infty)$ and $(0, \infty)$ respectively, $g(t) = t^r$ is operator convex on $(0, \infty)$ for $-1 \leq r \leq 0$ and $1 \leq r \leq 2$, from Theorems 1.6, 1.7 we get the following corollary.

Corollary 1.8 *Let Φ be a unital positive linear map from M_m to M_n. Then*

$$\Phi(A^r) \leq \Phi(A)^r, \quad A \geq 0, \quad 0 < r \leq 1;$$

$$\Phi(A^r) \geq \Phi(A)^r, \quad A > 0, \quad -1 \leq r \leq 0 \ \text{or} \ 1 \leq r \leq 2;$$

$$\Phi(\log A) \leq \log(\Phi(A)), \quad A > 0.$$

Given $A = (a_{ij}), B = (b_{ij}) \in M_n$, the *Hadamard product* of A and B is defined as the entry-wise product: $A \circ B \equiv (a_{ij}b_{ij}) \in M_n$. For this topic see

[52, Chapter 5]. We denote by $A[\alpha]$ the principal submatrix of A indexed by α. The following simple observation is very useful.

Lemma 1.9 *For any* $A, B \in M_n$, $A \circ B = (A \otimes B)[\alpha]$ *where* $\alpha = \{1, n + 2, 2n + 3, \ldots, n^2\}$. *Consequently there is a unital positive linear map* Φ *from* M_{n^2} *to* M_n *such that* $\Phi(A \otimes B) = A \circ B$ *for all* $A, B \in M_n$.

As an illustration of the usefulness of this lemma, consider the following reasoning: If $A, B \geq 0$, then evidently $A \otimes B \geq 0$. Since $A \circ B$ is a principal submatrix of $A \otimes B$, $A \circ B \geq 0$. Similarly $A \circ B > 0$ for the case when both A and B are positive definite. In other words, the Hadamard product of positive semidefinite (definite) matrices is positive semidefinite (definite). This important fact is known as the Schur product theorem.

Corollary 1.10

$$A^r \circ B^r \leq (A \circ B)^r, \quad A, B \geq 0, \quad 0 < r \leq 1; \tag{1.13}$$

$$A^r \circ B^r \geq (A \circ B)^r, \quad A, B > 0, \quad -1 \leq r \leq 0 \text{ or } 1 \leq r \leq 2; \tag{1.14}$$

$$(\log A + \log B) \circ I \leq \log(A \circ B), \quad A, B > 0. \tag{1.15}$$

Proof. This is an application of Corollary 1.8 with A there replaced by $A \otimes B$ and Φ being defined in Lemma 1.9.

For (1.13) and (1.14) just use the fact that $(A \otimes B)^t = A^t \otimes B^t$ for real number t. See [52] for properties of the Kronecker product.

For (1.15) we have

$$\log(A \otimes B) = \frac{d}{dt}(A \otimes B)^t|_{t=0} = \frac{d}{dt}(A^t \otimes B^t)|_{t=0}$$
$$= (\log A) \otimes I + I \otimes (\log B).$$

This can also be seen by using the spectral decompositions of A and B. $\quad\square$

We remark that the inequality in (1.14) is also valid for $A, B \geq 0$ in the case $1 \leq r \leq 2$.

Given a positive integer k, let us denote the kth Hadamard power of $A = (a_{ij}) \in M_n$ by $A^{(k)} \equiv (a_{ij}^k) \in M_n$. Here are two interesting consequences of Corollary 1.10: For every positive integer k,

$$(A^r)^{(k)} \leq (A^{(k)})^r, \quad A \geq 0, \quad 0 < r \leq 1;$$

$$(A^r)^{(k)} \geq (A^{(k)})^r, \quad A > 0, \quad -1 \leq r \leq 0 \text{ or } 1 \leq r \leq 2.$$

Corollary 1.11 *For* $A, B \geq 0$, *the function* $f(t) = (A^t \circ B^t)^{1/t}$ *is increasing on* $[1, \infty)$, *i.e.,*

$$(A^s \circ B^s)^{1/s} \leq (A^t \circ B^t)^{1/t}, \quad 1 \leq s < t.$$

Proof. By Corollary 1.10 we have

$$A^s \circ B^s \leq (A^t \circ B^t)^{s/t}.$$

Then applying the Löwner-Heinz inequality with the power $1/s$ yields the conclusion. □

Let P_n be the set of positive semidefinite matrices in M_n. A map Ψ from $P_n \times P_n$ into P_m is called *jointly concave* if

$$\Psi(\lambda A + (1-\lambda)B, \lambda C + (1-\lambda)D) \geq \lambda\Psi(A, C) + (1-\lambda)\Psi(B, D)$$

for all $A, B, C, D \geq 0$ and $0 < \lambda < 1$.

For $A, B > 0$, the *parallel sum* of A and B is defined as

$$A : B = (A^{-1} + B^{-1})^{-1}.$$

Note that $A : B = A - A(A+B)^{-1}A$ and $2(A : B) = \{(A^{-1}+B^{-1})/2\}^{-1}$ is the *harmonic mean* of A, B. Since $A : B$ decreases as A, B decrease, we can define the parallel sum for general $A, B \geq 0$ by

$$A : B = \lim_{\epsilon \downarrow 0}\{(A + \epsilon I)^{-1} + (B + \epsilon I)^{-1}\}^{-1}.$$

Using Lemma 1.4 it is easy to verify that

$$A : B = \max\left\{X \geq 0 : \begin{bmatrix} A+B & A \\ A & A-X \end{bmatrix} \geq 0\right\}$$

where the maximum is with respect to the Löwner partial order. From this extremal representation it follows readily that the map $(A, B) \mapsto A : B$ is jointly concave.

Lemma 1.12 *For $0 < r < 1$ the map*

$$(A, B) \mapsto A^r \circ B^{1-r}$$

is jointly concave in $A, B \geq 0$.

Proof. It suffices to prove that the map $(A, B) \mapsto A^r \otimes B^{1-r}$ is jointly concave in $A, B \geq 0$, since then the assertion will follow via Lemma 1.9.

We may assume $B > 0$. Using $A^r \otimes B^{1-r} = (A \otimes B^{-1})^r(I \otimes B)$ and the integral representation

$$t^r = \frac{\sin r\pi}{\pi} \int_0^\infty \frac{s^{r-1}t}{s+t}ds \quad (0 < r < 1)$$

we get

$$A^r \otimes B^{1-r} = \frac{\sin r\pi}{\pi} \int_0^\infty s^{r-1}(A \otimes B^{-1})(A \otimes B^{-1} + sI \otimes I)^{-1}(I \otimes B)ds.$$

Since $A \otimes B^{-1}$ and $I \otimes B$ commute, it is easy to see that

$$(A \otimes B^{-1})(A \otimes B^{-1} + sI \otimes I)^{-1}(I \otimes B) = (s^{-1}A \otimes I) : (I \otimes B).$$

We know that the parallel sum is jointly concave. Thus the integrand above is also jointly concave, and so is $A^r \otimes B^{1-r}$. This completes the proof. □

Corollary 1.13 *For $A, B, C, D \geq 0$ and $p, q > 1$ with $1/p + 1/q = 1$,*

$$A \circ B + C \circ D \leq (A^p + C^p)^{1/p} \circ (B^q + D^q)^{1/q}.$$

Proof. This is just the mid-point joint concavity case $\lambda = 1/2$ of Lemma 1.12 with $r = 1/p$. □

Let $f(x)$ be a real-valued differentiable function defined on some real interval. We denote by $\Delta f(x, y) \equiv [f(x) - f(y)]/(x - y)$ the difference quotient where $\Delta f(x, x) \equiv f'(x)$.

Let $H(t) \in M_n$ be a family of Hermitian matrices for t in an open real interval (a, b) and suppose the eigenvalues of $H(t)$ are contained in some open real interval Ω for all $t \in (a, b)$. Let $H(t) = U(t)\Lambda(t)U(t)^*$ be the spectral decomposition with $U(t)$ unitary and $\Lambda(t) = \mathrm{diag}(\lambda_1(t), \ldots, \lambda_n(t))$. Assume that $H(t)$ is continuously differentiable on (a, b) and $f : \Omega \to \mathbb{R}$ is a continuously differentiable function. Then it is known [52, Theorem 6.6.30] that $f(H(t))$ is continuously differentiable and

$$\frac{d}{dt}f(H(t)) = U(t)\{[\Delta f(\lambda_i(t), \lambda_j(t))] \circ [U(t)^* H'(t)U(t)]\}U(t)^*.$$

Theorem 1.14 *For $A, B \geq 0$ and $p, q > 1$ with $1/p + 1/q = 1$,*

$$A \circ B \leq (A^p \circ I)^{1/p}(B^q \circ I)^{1/q}.$$

Proof. Denote

$$C \equiv (A^p \circ I)^{1/p} \equiv \mathrm{diag}(\lambda_1, \ldots, \lambda_n),$$

$$D \equiv (B^q \circ I)^{1/q} \equiv \mathrm{diag}(\mu_1, \ldots, \mu_n).$$

By continuity we may assume that $\lambda_i \neq \lambda_j$ and $\mu_i \neq \mu_j$ for $i \neq j$.

Using the above differential formula we compute

$$\frac{d}{dt}(C^p + tA^p)^{1/p}\Big|_{t=0} = X \circ A^p$$

and

$$\frac{d}{dt}(D^q + tB^q)^{1/q}\bigg|_{t=0} = Y \circ B^q$$

where $X = (x_{ij})$ and $Y = (y_{ij})$ are defined by

$$x_{ij} = (\lambda_i - \lambda_j)(\lambda_i^p - \lambda_j^p)^{-1} \text{ for } i \neq j \text{ and } x_{ii} = p^{-1}\lambda_i^{1-p},$$

$$y_{ij} = (\mu_i - \mu_j)(\mu_i^q - \mu_j^q)^{-1} \text{ for } i \neq j \text{ and } y_{ii} = q^{-1}\mu_i^{1-q}.$$

By Corollary 1.13

$$C \circ D + tA \circ B \leq (C^p + tA^p)^{1/p} \circ (D^q + tB^q)^{1/q}$$

for any $t \geq 0$. Therefore, via differentiation at $t = 0$ we have

$$\begin{aligned}
A \circ B &\leq \frac{d}{dt}(C^p + tA^p)^{1/p} \circ (D^q + tB^q)^{1/q}|_{t=0} \\
&= X \circ A^p \circ D + C \circ Y \circ B^q \\
&= (X \circ I)(A^p \circ I)D + C(Y \circ I)(B^q \circ I) \\
&= p^{-1}C^{1-p}(A^p \circ I)D + q^{-1}CD^{1-q}(B^q \circ I) \\
&= (A^p \circ I)^{1/p}(B^q \circ I)^{1/q}.
\end{aligned}$$

This completes the proof.   □

We will need the following result in the next section and in Chapter 3. See [17] for a proof.

Theorem 1.15 *Let f be an operator monotone function on $[0, \infty)$, g an operator convex function on $[0, \infty)$ with $g(0) \leq 0$. Then for every contraction C, i.e., $\|C\|_\infty \leq 1$ and every $A \geq 0$,*

$$f(C^*AC) \geq C^*f(A)C, \tag{1.16}$$

$$g(C^*AC) \leq C^*g(A)C. \tag{1.17}$$

Notes and References. As already remarked, Theorem 1.3 is part of the Löwner theory. The inequality (1.16) in Theorem 1.15 is due to F. Hansen [43] while the inequality (1.17) is proved by F. Hansen and G. K. Pedersen [44]. All other results in this section are due to T. Ando [3, 8].

1.3 Inequalities for Matrix Powers

The purpose of this section is to prove the following result.

Theorem 1.16 *If $A \geq B \geq 0$ then*

$$(B^r A^p B^r)^{1/q} \geq B^{(p+2r)/q} \tag{1.18}$$

and

$$A^{(p+2r)/q} \geq (A^r B^p A^r)^{1/q} \tag{1.19}$$

for $r \geq 0$, $p \geq 0$, $q \geq 1$ with $(1 + 2r)q \geq p + 2r$.

Proof. We abbreviate "the Löwner-Heinz inequality" to LH, and first prove (1.18).

If $0 \leq p < 1$, then by LH, $A^p \geq B^p$ and hence $B^r A^p B^r \geq B^{p+2r}$. Applying LH again with the power $1/q$ gives (1.18).

Next we consider the case $p \geq 1$. It suffices to prove

$$(B^r A^p B^r)^{(1+2r)/(p+2r)} \geq B^{1+2r}$$

for $r \geq 0$, $p \geq 1$, since by assumption $q \geq (p + 2r)/(1 + 2r)$, and then (1.18) follows from this inequality via LH. Let us introduce t to write the above inequality as

$$(B^r A^p B^r)^t \geq B^{1+2r}, \quad t = \frac{1 + 2r}{p + 2r}. \tag{1.20}$$

Note that $0 < t \leq 1$, as $p \geq 1$. We will show (1.20) by induction on $k = 0, 1, 2, \ldots$ for the intervals $(2^{k-1} - 1/2, 2^k - 1/2]$ containing r. Since $(0, \infty) = \cup_{k=0}^{\infty}(2^{k-1} - 1/2, 2^k - 1/2]$, (1.20) is proved.

By the standard continuity argument, we may and do assume that A, B are positive definite. First consider the case $k = 0$, i.e., $0 < r \leq 1/2$. By LH $A^{2r} \geq B^{2r}$ and hence $B^r A^{-2r} B^r \leq I$, which means that $A^{-r} B^r$ is a contraction. Applying (1.16) in Theorem 1.15 with $f(x) = x^t$ yields

$$\begin{aligned}
(B^r A^p B^r)^t &= [(A^{-r} B^r)^* A^{p+2r}(A^{-r} B^r)]^t \\
&\geq (A^{-r} B^r)^* A^{(p+2r)t}(A^{-r} B^r) \\
&= B^r A B^r \geq B^{1+2r},
\end{aligned}$$

proving (1.20) for the case $k = 0$.

Now suppose that (1.20) is true for $r \in (2^{k-1} - 1/2, 2^k - 1/2]$. Denote $A_1 = (B^r A^p B^r)^t$, $B_1 = B^{1+2r}$. Then our assumption is

$$A_1 \geq B_1 \quad \text{with} \quad t = \frac{1 + 2r}{p + 2r}.$$

Since $p_1 \equiv 1/t \geq 1$, apply the already proved case $r_1 \equiv 1/2$ to $A_1 \geq B_1$ to get

$$(B_1^{r_1} A_1^{p_1} B_1^{r_1})^{t_1} \geq B_1^{1+2r_1}, \quad t_1 \equiv \frac{1 + 2r_1}{p_1 + 2r_1}. \tag{1.21}$$

Note that $t_1 = \frac{2+4r}{p+4r+1}$. Denote $s = 2r + 1/2$. We have $s \in (2^k - 1/2, 2^{k+1} - 1/2]$. Then explicitly (1.21) is

$$(B^s A^p B^s)^{t_1} \geq B^{1+2s}, \quad t_1 = \frac{1+2s}{p+2s},$$

which shows that (1.20) holds for $r \in (2^k - 1/2, 2^{k+1} - 1/2]$. This completes the inductive argument and (1.18) is proved.

$A \geq B > 0$ implies $B^{-1} \geq A^{-1} > 0$. In (1.18) replacing A, B by B^{-1}, A^{-1} respectively yields (1.19).   □

The case $q = p \geq 1$ of Theorem 1.16 is the following

Corollary 1.17 *If* $A \geq B \geq 0$ *then*

$$(B^r A^p B^r)^{1/p} \geq B^{(p+2r)/p},$$

$$A^{(p+2r)/p} \geq (A^r B^p A^r)^{1/p}$$

for all $r \geq 0$ *and* $p \geq 1$.

A still more special case is the next

Corollary 1.18 *If* $A \geq B \geq 0$ *then*

$$(BA^2B)^{1/2} \geq B^2 \quad \text{and} \quad A^2 \geq (AB^2A)^{1/2}.$$

At first glance, Corollary 1.18 (and hence Theorem 1.16) is strange: For positive numbers $a \geq b$, we have $a^2 \geq (ba^2b)^{1/2} \geq b^2$. We know the matrix analog that $A \geq B \geq 0$ implies $A^2 \geq B^2$ is false, but Corollary 1.18 asserts that the matrix analog of the stronger inequality $(ba^2b)^{1/2} \geq b^2$ holds.

This example shows that when we move from the commutative world to the noncommutative one, direct generalizations may be false, but a judicious modification may be true.

Notes and References. Corollary 1.18 is a conjecture of N. N. Chan and M. K. Kwong [29]. T. Furuta [38] solved this conjecture by proving the more general Theorem 1.16. See [39] for a related result.

1.4 Block Matrix Techniques

In the proof of Lemma 1.5 we have seen that block matrix arguments are powerful. Here we give one more example. In later chapters we will employ other types of block matrix techniques.

Theorem 1.19 *Let* A, B, X, Y *be matrices with* A, B *positive definite and* X, Y *arbitrary. Then*

$$(X^* A^{-1} X) \circ (Y^* B^{-1} Y) \geq (X \circ Y)^* (A \circ B)^{-1} (X \circ Y), \tag{1.22}$$

$$X^*A^{-1}X + Y^*B^{-1}Y \geq (X+Y)^*(A+B)^{-1}(X+Y). \qquad (1.23)$$

Proof. By Lemma 1.4 we have

$$\begin{bmatrix} A & X \\ X^* & X^*A^{-1}X \end{bmatrix} \geq 0, \quad \begin{bmatrix} B & Y \\ Y^* & Y^*B^{-1}Y \end{bmatrix} \geq 0.$$

Applying the Schur product theorem gives

$$\begin{bmatrix} A \circ B & X \circ Y \\ (X \circ Y)^* & (X^*A^{-1}X) \circ (Y^*B^{-1}Y) \end{bmatrix} \geq 0. \qquad (1.24)$$

Applying Lemma 1.4 again in another direction to (1.24) yields (1.22).
The inequality (1.23) is proved in a similar way.   □

Now let us consider some useful special cases of this theorem. Choosing $A = B = I$ and $X = Y = I$ in (1.22) respectively we get

Corollary 1.20 *For any* X, Y *and positive definite* A, B

$$(X^*X) \circ (Y^*Y) \geq (X \circ Y)^*(X \circ Y), \qquad (1.25)$$

$$A^{-1} \circ B^{-1} \geq (A \circ B)^{-1}. \qquad (1.26)$$

In (1.26) setting $B = A^{-1}$ we get $A \circ A^{-1} \geq (A \circ A^{-1})^{-1}$ or equivalently

$$A \circ A^{-1} \geq I, \quad \text{for} \quad A > 0. \qquad (1.27)$$

(1.27) is a well-known inequality due to M. Fiedler.
Note that both (1.22) and (1.23) can be extended to the case of arbitrarily finite number of matrices by the same proof. For instance we have

$$\sum_1^k X_j^* A_j^{-1} X_j \geq \left(\sum_1^k X_j \right)^* \left(\sum_1^k A_j \right)^{-1} \left(\sum_1^k X_j \right)$$

for any X_j and $A_j > 0$, $j = 1, \ldots, k$, two special cases of which are particularly interesting:

$$k \sum_1^k X_j^* X_j \geq \left(\sum_1^k X_j \right)^* \left(\sum_1^k X_j \right),$$

$$\sum_1^k A_j^{-1} \geq k^2 \left(\sum_1^k A_j \right)^{-1}, \quad \text{each } A_j > 0.$$

We record the following fact for later use. Compare it with Lemma 1.4.

Lemma 1.21

$$\begin{bmatrix} A & B \\ B^* & C \end{bmatrix} \geq 0 \tag{1.28}$$

if and only if $A \geq 0$, $C \geq 0$ and there exists a contraction W such that $B = A^{1/2}WC^{1/2}$.

Proof. Recall the fact that

$$\begin{bmatrix} I & X \\ X^* & I \end{bmatrix} \geq 0$$

if and only if X is a contraction. The "if" part is easily checked.

Conversely suppose we have (1.28). First consider the case when $A > 0$, $C > 0$. Then

$$\begin{bmatrix} I & A^{-1/2}BC^{-1/2} \\ (A^{-1/2}BC^{-1/2})^* & I \end{bmatrix}$$

$$= \begin{bmatrix} A^{-1/2} & 0 \\ 0 & C^{-1/2} \end{bmatrix} \begin{bmatrix} A & B \\ B^* & C \end{bmatrix} \begin{bmatrix} A^{-1/2} & 0 \\ 0 & C^{-1/2} \end{bmatrix} \geq 0.$$

Thus $W \equiv A^{-1/2}BC^{-1/2}$ is a contraction and $B = A^{1/2}WC^{1/2}$.

Next for the general case we have

$$\begin{bmatrix} A + m^{-1}I & B \\ B^* & C + m^{-1}I \end{bmatrix} \geq 0$$

for any positive integer m. By what we have just proved, for each m there exists a contraction W_m such that

$$B = (A + m^{-1}I)^{1/2}W_m(C + m^{-1}I)^{1/2}. \tag{1.29}$$

Since the space of matrices of a given order is finite-dimensional, the unit ball of any norm is compact (here we are using the spectral norm). It follows that there is a convergent subsequence of $\{W_m\}_{m=1}^{\infty}$, say, $\lim_{k\to\infty} W_{m_k} = W$. Of course W is a contraction. Taking the limit $k \to \infty$ in (1.29) we obtain $B = A^{1/2}WC^{1/2}$. $\qquad\square$

Notes and References. Except Lemma 1.21, this section is taken from [86]. Note that Theorem 1.19 remains true for rectangular matrices X, Y. The inequality (1.22) is also proved independently in [84].

2. Majorization and Eigenvalues

Majorization is one of the most powerful techniques for deriving inequalities. We first introduce in Section 2.1 the concepts of four kinds of majorizations, give some examples related to matrices, and present several basic majorization principles. Then in Section 2.2 we prove two theorems on eigenvalues of the Hadamard product of positive semidefinite matrices, which generalize Oppenheim's classical inequalities.

2.1 Majorizations

Given a real vector $x = (x_1, x_2, \ldots, x_n) \in \mathbb{R}^n$, we rearrange its components as $x_{[1]} \geq x_{[2]} \geq \cdots \geq x_{[n]}$.

Definition. For $x = (x_1, \ldots, x_n)$, $y = (y_1, \ldots, y_n) \in \mathbb{R}^n$, if

$$\sum_{i=1}^{k} x_{[i]} \leq \sum_{i=1}^{k} y_{[i]}, \quad k = 1, 2, \ldots, n$$

then we say that x is *weakly majorized* by y and denote $x \prec_w y$. If in addition to $x \prec_w y$, $\sum_{i=1}^{n} x_i = \sum_{i=1}^{n} y_i$ holds, then we say that x is *majorized* by y and denote $x \prec y$.

For example, if each $a_i \geq 0$, $\sum_1^n a_i = 1$ then

$$(\frac{1}{n}, \ldots, \frac{1}{n}) \prec (a_1, \ldots, a_n) \prec (1, 0, \ldots, 0).$$

There is a useful characterization of majorization. We call a matrix *nonnegative* if all its entries are nonnegative real numbers. A nonnegative matrix is called *doubly stochastic* if all its row and column sums are one. Let $x, y \in \mathbb{R}^n$. The Hardy-Littlewood-Pólya theorem ([17, Theorem II.1.10] or [72, p.22]) asserts that $x \prec y$ if and only if there exists a doubly stochastic matrix A such that $x = Ay$. Here we regard vectors as column vectors, i.e., $n \times 1$ matrices.

By this characterization we readily get the following well-known theorem of Schur via the spectral decomposition of Hermitian matrices.

Theorem 2.1 *If H is a Hermitian matrix with diagonal entries $h_1, \ldots, h_n$ and eigenvalues $\lambda_1, \ldots, \lambda_n$ then*

$$(h_1, \ldots, h_n) \prec (\lambda_1, \ldots, \lambda_n). \tag{2.1}$$

In the sequel, if the eigenvalues of a matrix H are all real, we will always arrange them in decreasing order: $\lambda_1(H) \geq \lambda_2(H) \geq \cdots \geq \lambda_n(H)$ and denote $\lambda(H) \equiv (\lambda_1(H), \ldots, \lambda_n(H))$. If G, H are Hermitian matrices and $\lambda(G) \prec \lambda(H)$, we simply write $G \prec H$. Similarly we write $G \prec_w H$ to indicate $\lambda(G) \prec_w \lambda(H)$. For example, Theorem 2.1 can be written as

$$H \circ I \prec H \quad \text{for Hermitian } H. \tag{2.2}$$

The next two majorization principles [72, p.115 and 116] are of primary importance. Here we assume that the functions $f(t)$, $g(t)$ are defined on some interval containing the components of $x = (x_1, \ldots, x_n)$ and $y = (y_1, \ldots, y_n)$.

Theorem 2.2 *Let $f(t)$ be a convex function. Then*

$$x \prec y \text{ implies } (f(x_1), \ldots, f(x_n)) \prec_w (f(y_1), \ldots, f(y_n)).$$

Theorem 2.3 *Let $g(t)$ be an increasing convex function. Then*

$$x \prec_w y \text{ implies } (g(x_1), \ldots, g(x_n)) \prec_w (g(y_1), \ldots, g(y_n)).$$

To illustrate the effect of Theorem 2.2, suppose in Theorem 2.1 $H > 0$ and without loss of generality, $h_1 \geq \cdots \geq h_n$, $\lambda_1 \geq \cdots \geq \lambda_n$. Then apply Theorem 2.2 with $f(t) = -\log t$ to the majorization (2.1) to get

$$\prod_{i=k}^{n} h_i \geq \prod_{i=k}^{n} \lambda_i, \quad k = 1, 2, \ldots, n. \tag{2.3}$$

Of course the condition $H > 0$ can be relaxed to $H \geq 0$ by continuity. Note that the special case $k = 1$ of (2.3) says $\det H \leq \prod_1^n h_i$, which is called the Hadamard inequality.

Definition. Let the components of $x = (x_1, \ldots, x_n)$ and $y = (y_1, \ldots, y_n)$ be nonnegative. If

$$\prod_{i=1}^{k} x_{[i]} \leq \prod_{i=1}^{k} y_{[i]}, \quad k = 1, 2, \ldots, n$$

then we say that x is *weakly log-majorized* by y and denote $x \prec_{wlog} y$. If in addition to $x \prec_{wlog} y$, $\prod_{i=1}^{n} x_i = \prod_{i=1}^{n} y_i$ holds, then we say that x is *log-majorized* by y and denote $x \prec_{log} y$.

The absolute value of a matrix A is, by definition, $|A| \equiv (A^*A)^{1/2}$. The *singular values* of A are defined to be the eigenvalues of $|A|$. Thus the singular values of A are the nonnegative square roots of the eigenvalues of A^*A. For positive semidefinite matrices, singular values and eigenvalues coincide.

Throughout we arrange the singular values of A in decreasing order: $s_1(A) \geq \cdots \geq s_n(A)$ and denote $s(A) \equiv (s_1(A), \ldots, s_n(A))$. Note that the spectral norm of A, $\|A\|_\infty$, is equal to $s_1(A)$.

Let us write $\{x_i\}$ for a vector $(x_1, \ldots, x_n)$. In matrix theory there are the following three basic majorization relations [52].

Theorem 2.4 (H. Weyl) *Let* $\lambda_1(A), \ldots, \lambda_n(A)$ *be the eigenvalues of a matrix* A *ordered so that* $|\lambda_1(A)| \geq \cdots \geq |\lambda_n(A)|$. *Then*

$$\{|\lambda_i(A)|\} \prec_{log} s(A).$$

Theorem 2.5 (A. Horn) *For any matrices* A, B

$$s(AB) \prec_{log} \{s_i(A)s_i(B)\}.$$

Theorem 2.6 *For any matrices* A, B

$$s(A \circ B) \prec_w \{s_i(A)s_i(B)\}.$$

Note that the eigenvalues of the product of two positive semidefinite matrices are nonnegative, since $\lambda(AB) = \lambda(A^{1/2}BA^{1/2})$.

If A, B are positive semidefinite, then by Theorems 2.4 and 2.5

$$\lambda(AB) \prec_{log} s(AB) \prec_{log} \{\lambda_i(A)\lambda_i(B)\}.$$

Therefore

$$A, B \geq 0 \ \Rightarrow \ \lambda(AB) \prec_{log} \{\lambda_i(A)\lambda_i(B)\}. \tag{2.4}$$

We remark that for nonnegative vectors, weak log-majorization is stronger than weak majorization, which follows from the case $g(t) = e^t$ of Theorem 2.3. We record this as

Theorem 2.7 *Let the components of* $x, y \in \mathbb{R}^n$ *be nonnegative. Then*

$$x \prec_{wlog} y \quad \text{implies} \quad x \prec_w y$$

Applying Theorem 2.7 to Theorems 2.4 and 2.5 we get the following

Corollary 2.8 *For any* $A, B \in M_n$

$$|\operatorname{tr} A| \le \sum_{i=1}^{n} s_i(A),$$

$$s(AB) \prec_w \{s_i(A)s_i(B)\}.$$

The next result [17, proof of Theorem IX.2.9] is very useful.

Theorem 2.9 *Let $A, B \ge 0$. If $0 < s < t$ then*

$$\{\lambda_j^{1/s}(A^s B^s)\} \prec_{log} \{\lambda_j^{1/t}(A^t B^t)\}.$$

By this theorem, if $A, B \ge 0$ and $m \ge 1$ then

$$\{\lambda_j^m(AB)\} \prec_{log} \{\lambda_j(A^m B^m)\}.$$

Since weak log-majorization implies weak majorization, we have

$$\{\lambda_j^m(AB)\} \prec_w \{\lambda_j(A^m B^m)\}.$$

Now if m is a positive integer, $\lambda_j^m(AB) = \lambda_j[(AB)^m]$ and hence

$$\{\lambda_j[(AB)^m]\} \prec_w \{\lambda_j(A^m B^m)\}.$$

In particular we have

$$\operatorname{tr}(AB)^m \le \operatorname{tr} A^m B^m.$$

Theorem 2.10 *Let G, H be Hermitian. Then*

$$\lambda(e^{G+H}) \prec_{log} \lambda(e^G e^H).$$

Proof. Let $G, H \in M_n$ and $1 \le k \le n$ be fixed. By the spectral mapping theorem and Theorem 2.9, for any positive integer m

$$\prod_{j=1}^{k} \lambda_j[(e^{\frac{G}{m}} e^{\frac{H}{m}})^m] = \prod_{j=1}^{k} \lambda_j^m(e^{\frac{G}{m}} e^{\frac{H}{m}}) \le \prod_{j=1}^{k} \lambda_j(e^G e^H). \qquad (2.5)$$

The Lie product formula [17, p.254] says

$$\lim_{m \to \infty} (e^{\frac{X}{m}} e^{\frac{Y}{m}})^m = e^{X+Y}$$

for any two matrices X, Y. Thus letting $m \to \infty$ in (2.5) yields

$$\lambda(e^{G+H}) \prec_{wlog} \lambda(e^G e^H).$$

Finally note that $\det e^{G+H} = \det(e^G e^H)$. This completes the proof.   □

Note that Theorem 2.10 strengthens the Golden-Thompson inequality:

$$\operatorname{tr} e^{G+H} \leq \operatorname{tr} e^G e^H$$

for Hermitian G, H.

From the minimax characterization of eigenvalues of a Hermitian matrix [17] it follows immediately that $A \geq B$ implies $\lambda_j(A) \geq \lambda_j(B)$ for each j. This fact will be repeatedly used in the sequel.

Notes and References. For a more detailed treatment of the majorization theory see [72, 5, 17, 52]. For the topic of log-majorization see [46].

2.2 Eigenvalues of Hadamard Products

Let $A, B = (b_{ij}) \in M_n$ be positive semidefinite. Oppenheim's inequality states that

$$\det(A \circ B) \geq (\det A) \prod_{i=1}^{n} b_{ii}. \tag{2.6}$$

By Hadamard's inequality, (2.6) implies

$$\det(A \circ B) \geq \det(AB). \tag{2.7}$$

We first give a generalization of the determinantal inequality (2.7), whose proof uses several results we obtained earlier, and then we generalize (2.6).

Let $x = \{x_i\}, y = \{y_i\} \in \mathbb{R}^n$ with $x_1 \geq \cdots \geq x_n, y_1 \geq \cdots \geq y_n$. Observe that if $x \prec y$ then

$$\sum_{i=k}^{n} x_i \geq \sum_{i=k}^{n} y_i, \quad k = 1, \ldots, n.$$

Also if the components of x and y are positive and $x \prec_{log} y$ then

$$\prod_{i=k}^{n} x_i \geq \prod_{i=k}^{n} y_i, \quad k = 1, \ldots, n.$$

Theorem 2.11 *Let $A, B \in M_n$ be positive definite. Then*

$$\prod_{j=k}^{n} \lambda_j(A \circ B) \geq \prod_{j=k}^{n} \lambda_j(AB), \quad k = 1, 2, \ldots, n. \tag{2.8}$$

Proof. By (1.15) in Corollary 1.10 we have

$$\log(A \circ B) \geq (\log A + \log B) \circ I$$

which implies

$$\log[\prod_{j=k}^{n} \lambda_j(A \circ B)] = \sum_{j=k}^{n} \lambda_j[\log(A \circ B)]$$

$$\geq \sum_{j=k}^{n} \lambda_j[(\log A + \log B) \circ I]$$

for $k = 1, 2, \ldots, n$. According to Schur's theorem (see (2.2))

$$(\log A + \log B) \circ I \prec \log A + \log B,$$

from which it follows that

$$\sum_{j=k}^{n} \lambda_j[(\log A + \log B) \circ I] \geq \sum_{j=k}^{n} \lambda_j(\log A + \log B)$$

for $k = 1, 2, \ldots, n$.

On the other hand, in Theorem 2.10 setting $G = \log A$, $H = \log B$ we have

$$\lambda(e^{\log A + \log B}) \prec_{log} \lambda(AB).$$

Since $\lambda_j(e^{\log A + \log B}) = e^{\lambda_j(\log A + \log B)}$, this log-majorization is equivalent to the majorization

$$\lambda(\log A + \log B) \prec \{\log \lambda_j(AB)\}.$$

But $\log \lambda_j(AB) = \log \lambda_j(A^{1/2} B A^{1/2}) = \lambda_j[\log(A^{1/2} B A^{1/2})]$, so

$$\log A + \log B \prec \log(A^{1/2} B A^{1/2}).$$

Hence

$$\sum_{j=k}^{n} \lambda_j(\log A + \log B) \geq \sum_{j=k}^{n} \lambda_j[\log(A^{1/2} B A^{1/2})]$$

$$= \log[\prod_{j=k}^{n} \lambda_j(AB)]$$

for $k = 1, 2, \ldots, n$. Combining the above three inequalities involving eigenvalues we get

$$\prod_{j=k}^{n} \lambda_j(A \circ B) \geq \prod_{j=k}^{n} \lambda_j(AB), \quad k = 1, 2, \ldots, n$$

which establishes the theorem. $\square$

Denote by G^T the transpose of a matrix G. Since for $B > 0$ $\log B^T = (\log B)^T$, we have

$$(\log B^T) \circ I = (\log B)^T \circ I = (\log B) \circ I.$$

Therefore in the above proof we can replace $(\log A + \log B) \circ I$ by $(\log A + \log B^T) \circ I$. Thus we get the following

Theorem 2.12 *Let* $A, B \in M_n$ *be positive definite. Then*

$$\prod_{j=k}^{n} \lambda_j(A \circ B) \geq \prod_{j=k}^{n} \lambda_j(AB^T), \quad k = 1, 2, \ldots, n. \tag{2.9}$$

Note that the special case $k = 1$ of (2.8) is the inequality (2.7). For $A, B > 0$, by the log-majorization (2.4) we have

$$\prod_{j=k}^{n} \lambda_j(AB) \geq \prod_{j=k}^{n} \lambda_j(A)\lambda_j(B), \quad k = 1, 2, \ldots, n. \tag{2.10}$$

Combining (2.8) and (2.10) we get the following

Corollary 2.13 *Let* $A, B \in M_n$ *be positive definite. Then*

$$\prod_{j=k}^{n} \lambda_j(A \circ B) \geq \prod_{j=k}^{n} \lambda_j(A)\lambda_j(B), \quad k = 1, 2, \ldots, n.$$

Next we give a generalization of Oppenheim's inequality (2.6).

A linear map $\Phi : M_n \to M_n$ is said to be *doubly stochastic* if it is positive ($A \geq 0 \Rightarrow \Phi(A) \geq 0$), unital ($\Phi(I) = I$) and trace-preserving ($\operatorname{tr}\Phi(A) = \operatorname{tr}A$ for all $A \in M_n$). Since every Hermitian matrix can be written as a difference of two positive semidefinite matrices: $H = (|H| + H)/2 - (|H| - H)/2$, a positive linear map necessarily preserves the set of Hermitian matrices.

The Frobenius inner product on M_n is $\langle A, B \rangle \equiv \operatorname{tr}AB^*$.

Lemma 2.14 *Let* $A \in M_n$ *be Hermitian and* $\Phi : M_n \to M_n$ *be a doubly stochastic map. Then*

$$\Phi(A) \prec A.$$

Proof. Let

$$A = U\operatorname{diag}(x_1, \ldots, x_n)U^*, \quad \Phi(A) = W\operatorname{diag}(y_1, \ldots, y_n)W^*$$

be the spectral decompositions with U, W unitary. Define

$$\Psi(X) = W^*\Phi(UXU^*)W.$$

Then Ψ is again a doubly stochastic map and

$$\text{diag}(y_1, \ldots, y_n) = \Psi(\text{diag}(x_1, \ldots, x_n)). \tag{2.11}$$

Let $P_j \equiv e_j e_j^T$, the orthogonal projection to the one-dimensional subspace spanned by the jth standard basis vector e_j, $j = 1, \ldots, n$. Then (2.11) implies

$$(y_1, \ldots, y_n)^T = D(x_1, \ldots, x_n)^T \tag{2.12}$$

where $D = (d_{ij})$ with $d_{ij} = \langle \Psi(P_j), P_i \rangle$ for all i and j. Since Ψ is doubly stochastic, it is easy to verify that D is a doubly stochastic matrix. By the Hardy-Littlewood-Pólya theorem, the relation (2.12) implies

$$(y_1, \ldots, y_n) \prec (x_1, \ldots, x_n),$$

proving the lemma.   □

A positive semidefinite matrix with all diagonal entries 1 is called a *correlation matrix*.

Suppose C is a correlation matrix. Define $\Phi_C(X) = X \circ C$. Obviously Φ_C is a doubly stochastic map on M_n. Thus we have the following

Corollary 2.15 *If A is Hermitian and C is a correlation matrix. Then*

$$A \circ C \prec A.$$

Theorem 2.16 *Let $A, B \in M_n$ be positive definite and $\beta_1 \geq \cdots \geq \beta_n$ be a rearrangement of the diagonal entries of B. Then for every $1 \leq k \leq n$*

$$\prod_{j=k}^{n} \lambda_j(A \circ B) \geq \prod_{j=k}^{n} \lambda_j(A)\beta_j. \tag{2.13}$$

Proof. Let $B = (b_{ij})$. Define two matrices $C = (c_{ij})$ and $H = (h_{ij})$ by

$$c_{ij} = \frac{b_{ij}}{(b_{ii}b_{jj})^{1/2}}, \quad h_{ij} = (b_{ii}b_{jj})^{1/2}.$$

Then C is a correlation matrix and $B = H \circ C$. By Corollary 2.15

$$A \circ B = (A \circ H) \circ C \prec A \circ H.$$

Applying Theorem 2.2 with $f(t) = -\log t$ to this majorization yields

$$\prod_{j=k}^{n} \lambda_j(A \circ B) \geq \prod_{j=k}^{n} \lambda_j(A \circ H). \tag{2.14}$$

Note that $A \circ H = DAD$ where $D \equiv \mathrm{diag}(\sqrt{b_{11}}, \ldots, \sqrt{b_{nn}})$ and $\lambda_j(D^2) = \beta_j$, $j = 1, \ldots, n$. Thus $\lambda(A \circ H) = \lambda(DAD) = \lambda(AD^2)$. By the log-majorization (2.4) we have

$$\prod_{j=k}^{n} \lambda_j(A \circ H) = \prod_{j=k}^{n} \lambda_j(AD^2) \geq \prod_{j=k}^{n} \lambda_j(A)\lambda_j(D^2)$$

$$= \prod_{j=k}^{n} \lambda_j(A)\beta_j.$$

Combining (2.14) and the above inequality gives (2.13).   □

Oppenheim's inequality (2.6) corresponds to the special case $k = 1$ of (2.13). Also note that since $\prod_{j=k}^{n} \beta_j \geq \prod_{j=k}^{n} \lambda_j(B)$, Theorem 2.16 is stronger than Corollary 2.13.

We remark that in general Theorem 2.16 and Theorem 2.11 are not comparable. In fact, both $\lambda_n(AB) > \lambda_n(A)\beta_n$ and $\lambda_n(AB) < \lambda_n(A)\beta_n$ can occur: For $A = \mathrm{diag}(1, 2)$, $B = \mathrm{diag}(2, 1)$

$$\lambda_2(AB) - \lambda_2(A)\beta_2 = 1$$

while for

$$A = I_2, \ B = \begin{bmatrix} 2 & 1 \\ 1 & 2 \end{bmatrix}$$

$$\lambda_2(AB) - \lambda_2(A)\beta_2 = -1.$$

Notes and References. M. Fiedler [37] first proved the case $k = n$ of Theorem 2.12. Then C. R. Johnson and L. Elsner [58] proved the case $k = n$ of Theorem 2.11, and in [57] C. R. Johnson and R. B. Bapat conjectured the general inequalities (2.8) in Theorem 2.11. T. Ando [6] and G. Visick [82] independently solved this conjecture affirmatively. What we presented here is Ando's elegant proof. Theorem 2.12 is also proved in [6].

Corollary 2.13 is a conjecture of A. W. Marshall and I. Olkin [72, p.258]. R. B. Bapat and V. S. Sunder [12] solved this conjecture by proving the stronger Theorem 2.16. Corollary 2.15 is also due to them. Lemma 2.14 is due to T. Ando [5, Theorem 7.1].

3. Singular Values

Recall that the singular values of a matrix $A \in M_n$ are the eigenvalues of its absolute value $|A| \equiv (A^*A)^{1/2}$, and we have fixed the notation $s(A) \equiv (s_1(A), \ldots, s_n(A))$ with $s_1(A) \geq \cdots \geq s_n(A)$ for the singular values of A. Singular values are closely related to unitarily invariant norms, which are the theme of the next chapter. Singular value inequalities are weaker than Löwner partial order inequalities and stronger than unitarily invariant norm inequalities in the following sense:

$$|A| \leq |B| \Rightarrow s_j(A) \leq s_j(B), \text{ for each } j \Rightarrow \|A\| \leq \|B\|$$

for all unitarily invariant norms.

Note that singular values are unitarily invariant: $s(UAV) = s(A)$ for every A and all unitary U, V.

3.1 Matrix Young Inequalities

The most important case of the Young inequality says that if $1/p + 1/q = 1$ with $p, q > 1$ then

$$|ab| \leq \frac{|a|^p}{p} + \frac{|b|^q}{q} \quad \text{for} \quad a, b \in \mathbb{C}.$$

A direct matrix generalization would be

$$|AB| \leq \frac{|A|^p}{p} + \frac{|B|^q}{q}$$

which is false in general. If $A, B \geq 0$ is a commuting pair, however, $AB \geq 0$ and via a simultaneous unitary diagonalization [51, Corollary 4.5.18] it is clear that

$$AB \leq \frac{A^p}{p} + \frac{B^q}{q}$$

holds.

It turns out that the singular value version generalization of the Young inequality is true.

We will need the following special cases of Theorem 1.15.

Lemma 3.1 *Let Q be an orthogonal projection and $X \geq 0$. Then*

$$QX^r Q \leq (QXQ)^r \quad \text{if} \quad 0 < r \leq 1,$$

$$QX^r Q \geq (QXQ)^r \quad \text{if} \quad 1 \leq r \leq 2.$$

Theorem 3.2 *Let $p, q > 1$ with $1/p + 1/q = 1$. Then for any matrices $A, B \in M_n$*

$$s_j(AB^*) \leq s_j \left(\frac{|A|^p}{p} + \frac{|B|^q}{q} \right), \quad j = 1, 2, \ldots, n. \tag{3.1}$$

Proof. By considering the polar decompositions $A = V|A|$, $B = W|B|$ with V, W unitary, we see that it suffices to prove (3.1) for $A, B \geq 0$. Now we make this assumption.

Passing to eigenvalues, (3.1) means

$$\lambda_k((BA^2B)^{1/2}) \leq \lambda_k(A^p/p + B^q/q) \tag{3.2}$$

for each $1 \leq k \leq n$. Let us fix k and prove (3.2).

Since $\lambda_k((BA^2B)^{1/2}) = \lambda_k((AB^2A)^{1/2})$, by exchanging the roles of A and B if necessary, we may assume $1 < p \leq 2$, hence $2 \leq q < \infty$. Further by the standard continuity argument we may assume $B > 0$.

Here we regard matrices in M_n as linear operators on $\mathbb{C}^n$.

Write $\lambda \equiv \lambda_k((BA^2B)^{1/2})$ and denote by P the orthogonal projection (of rank k) to the spectral subspace spanned by the eigenvectors corresponding to $\lambda_j((BA^2B)^{1/2})$ for $j = 1, 2, \ldots, k$. Denote by Q the orthogonal projection (of rank k) to the subspace $\mathcal{M} \equiv \text{range}(B^{-1}P)$. In view of the minimax characterization of eigenvalues of a Hermitian matrix [17], for the inequality (3.2) it suffices to prove

$$\lambda Q \leq QA^pQ/p + QB^qQ/q. \tag{3.3}$$

By the definition of Q we have

$$QB^{-1}P = B^{-1}P \tag{3.4}$$

and that there exists G such that $Q = B^{-1}PG$, hence $BQ = PG$ and

$$PBQ = BQ. \tag{3.5}$$

Taking adjoints in (3.4) and (3.5) gives

$$PB^{-1}Q = PB^{-1} \tag{3.6}$$

and

$$QBP = QB. \tag{3.7}$$

By (3.4) and (3.7) we have

$$\begin{aligned}(QB^2Q) \cdot (B^{-1}PB^{-1}) &= QB^2 \cdot QB^{-1}PB^{-1}\\ &= QB^2 \cdot B^{-1}PB^{-1}\\ &= QBPB^{-1} = Q\end{aligned}$$

and similarly from (3.6) and (3.5) we get

$$(B^{-1}PB^{-1}) \cdot (QB^2Q) = Q.$$

These together mean that $B^{-1}PB^{-1}$ and QB^2Q map $\mathcal{M}$ onto itself, vanish on its orthogonal complement and are inverse to each other on $\mathcal{M}$.

By the definition of P we have

$$(BA^2B)^{1/2} \geq \lambda P$$

which implies, via commutativity of $(BA^2B)^{1/2}$ and P,

$$A^2 \geq \lambda^2 B^{-1}PB^{-1}.$$

Then by the Löwner-Heinz inequality (Theorem 1.1) with $r = p/2$ we get

$$A^p \geq \lambda^p \cdot (B^{-1}PB^{-1})^{p/2},$$

hence by (3.4) and (3.6)

$$QA^pQ \geq \lambda^p \cdot (B^{-1}PB^{-1})^{p/2}.$$

Since $B^{-1}PB^{-1}$ is the inverse of QB^2Q on $\mathcal{M}$, this means that on $\mathcal{M}$

$$QA^pQ \geq \lambda^p \cdot (QB^2Q)^{-p/2}. \tag{3.8}$$

To prove (3.3), let us first consider the case $2 \leq q \leq 4$. By Lemma 3.1 with $r = q/2$ we have

$$QB^qQ \geq (QB^2Q)^{q/2}. \tag{3.9}$$

Now it follows from (3.8) and (3.9) that on $\mathcal{M}$

$$QA^pQ/p + QB^qQ/q \geq \lambda^p \cdot (QB^2Q)^{-p/2}/p + (QB^2Q)^{q/2}/q.$$

In view of the Young inequality for the commuting pair, $\lambda \cdot (QB^2Q)^{-1/2}$ and $(QB^2Q)^{1/2}$, this implies

$$QA^pQ/p + QB^qQ/q \geq \lambda \cdot (QB^2Q)^{-1/2} \cdot (QB^2Q)^{1/2} = \lambda Q,$$

proving (3.3).

Let us next consider the case $4 < q < \infty$. Let $s = q/2$. Then $0 < 2/s < 1$ and $q/s = 2$. By Lemma 3.1 with $r = q/s$ we have

$$QB^qQ \geq (QB^sQ)^{q/s}. \tag{3.10}$$

On the other hand, by Lemma 3.1 with $r = 2/s$ we have

$$(QB^sQ)^{2/s} \geq QB^2Q,$$

and then by the Löwner-Heinz inequality with $r = p/2$

$$(QB^sQ)^{p/s} \geq (QB^2Q)^{p/2},$$

hence on $\mathcal{M}$

$$(QB^sQ)^{-p/s} \leq (QB^2Q)^{-p/2}. \tag{3.11}$$

Combining (3.8) and (3.11) yields

$$QA^pQ \geq \lambda^p \cdot (QB^sQ)^{-p/s}. \tag{3.12}$$

Now it follows from (3.12) and (3.10) that

$$QA^pQ/p + QB^qQ/q \geq \lambda^p \cdot (QB^sQ)^{-p/s}/p + (QB^sQ)^{q/s}/q.$$

In view of the Young inequality for the commuting pair, $\lambda \cdot (QB^sQ)^{-1/s}$ and $(QB^sQ)^{1/s}$, this implies

$$QA^pQ/p + QB^qQ/q \geq \lambda \cdot (QB^sQ)^{-1/s} \cdot (QB^sQ)^{1/s} = \lambda Q,$$

proving (3.3). This completes the proof. $\square$

The case $p = q = 2$ of Theorem 3.2 has the following form:

Corollary 3.3 *For any $X, Y \in M_n$*

$$2s_j(XY^*) \leq s_j(X^*X + Y^*Y), \quad j = 1, 2, \ldots, n. \tag{3.13}$$

The conclusion of Theorem 3.2 is equivalent to the statement that there exists a unitary matrix U, depending on A, B, such that

$$U|AB^*|U^* \leq \frac{|A|^p}{p} + \frac{|B|^q}{q}.$$

It seems natural to pose the following

Conjecture 3.4 *Let $A, B \in M_n$ be positive semidefinite and $0 \leq r \leq 1$. Then*

$$s_j(A^rB^{1-r} + A^{1-r}B^r) \leq s_j(A + B), \quad j = 1, 2, \ldots, n. \tag{3.14}$$

Observe that the special case $r = 1/2$ of (3.14) is just (3.13) while the cases $r = 0$, 1 are trivial.

Another related problem is the following

Question 3.5 *Let* $A, B \in M_n$ *be positive semidefinite. Is it true that*

$$s_j^{1/2}(AB) \leq \frac{1}{2}s_j(A + B), \quad j = 1, 2, \ldots, n? \tag{3.15}$$

Since the square function is operator convex on $\mathbb{R}$, i.e.,

$$\left(\frac{A + B}{2}\right)^2 \leq \frac{A^2 + B^2}{2},$$

the statement (3.15) is stronger than (3.13).

Notes and References. Theorem 3.2 is due to T. Ando [7]. Corollary 3.3 is due to R. Bhatia and F. Kittaneh [22]. Conjecture 3.4 is posed in [89] and Question 3.5 is in [24].

3.2 Singular Values of Hadamard Products

Given $A = (a_{ij}) \in M_n$, we denote the decreasingly ordered Euclidean row and column lengths of A by $r_1(A) \geq r_2(A) \geq \cdots \geq r_n(A)$ and $c_1(A) \geq c_2(A) \geq \cdots \geq c_n(A)$ respectively, i.e., $r_k(A)$ is the kth largest value of $(\sum_{j=1}^n |a_{ij}|^2)^{1/2}$, $i = 1, \ldots, n$ and $c_k(A)$ is the kth largest value of $(\sum_{i=1}^n |a_{ij}|^2)^{1/2}$, $j = 1, \ldots, n$.

The purpose of this section is to prove the following

Theorem 3.6 *For any* $A, B \in M_n$

$$s(A \circ B) \prec_w \{\min\{r_i(A), c_i(A)\}s_i(B)\}. \tag{3.16}$$

The proof of this theorem is divided into a series of lemmas. The first fact is easy to verify.

Lemma 3.7 *For any* $A, B, C \in M_n$, $(A \circ B)C$ *and* $(A \circ C^T)B^T$ *have the same main diagonal. In particular,*

$$\mathrm{tr}(A \circ B)C = \mathrm{tr}(A \circ C^T)B^T.$$

A matrix that has r singular values 1 and all other singular values 0 is called a rank r partial isometry.

Lemma 3.8 *For any $C \in M_n$ and $1 \le k \le n$, there is a rank k partial isometry $C_k \in M_n$ such that*

$$\sum_{i=1}^{k} s_i(C) = \operatorname{tr}(CC_k).$$

Proof. Let $C = U|C|$ be the polar decomposition with U unitary and $U|C|U^* = \sum_{i=1}^{n} s_i(C)P_i$ be the spectral decomposition where $P_1, \ldots, P_n$ are mutually orthogonal rank one projections. Then $C_k \equiv U^*(\sum_{i=1}^{k} P_i)$ does the job. □

Lemma 3.9 *Every $B \in M_n$ can be written as*

$$B = \sum_{i=1}^{n} \mu_i K_i$$

where each $\mu_i \ge 0$ and each K_i is a rank i partial isometry such that

$$\sum_{i=k}^{n} \mu_i = s_k(B), \quad k = 1, \ldots, n.$$

Proof. Let $B = U|B|$ be the polar decomposition with U unitary and $|B| = \sum_{i=1}^{n} s_i(B)P_i$ be the spectral decomposition where $P_1, \ldots, P_n$ are mutually orthogonal rank one projections. Then

$$\mu_i \equiv s_i(B) - s_{i+1}(B), \ i = 1, \ldots, n-1, \ \mu_n \equiv s_n(B)$$

and

$$K_i \equiv U \sum_{j=1}^{i} P_j, \quad i = 1, \ldots, n$$

satisfy the requirements. □

Lemma 3.10 *Let $A \in M_n$ and $\alpha_1 \ge \alpha_2 \ge \cdots \ge \alpha_n \ge 0$ be given. If*

$$s(A \circ B) \prec_w \{\alpha_i s_1(B)\} \quad \text{for all } B \in M_n, \tag{3.17}$$

then

$$s(A \circ B) \prec_w \{\alpha_i s_i(B)\} \quad \text{for all } B \in M_n. \tag{3.18}$$

Proof. Assume (3.17). We first show that if K_r, $K_t \in M_n$ are partial isometries with respective ranks r and t then

$$|\text{tr}(A \circ K_r)K_t| \leq \sum_{i=1}^{\min\{r,t\}} \alpha_i. \tag{3.19}$$

In view of Lemma 3.7, we may assume, without loss of generality, that $t \leq r$. Using Corollary 2.8 and the assumption (3.17) we compute

$$|\text{tr}(A \circ K_r)K_t| \leq \sum_{i=1}^{n} s_i[(A \circ K_r)K_t]$$

$$\leq \sum_{i=1}^{n} s_i(A \circ K_r)s_i(K_t)$$

$$= \sum_{i=1}^{t} s_i(A \circ K_r)$$

$$\leq \sum_{i=1}^{t} \alpha_i,$$

proving (3.19).

Next for any k with $1 \leq k \leq n$, by Lemma 3.8 there is a rank k partial isometry C_k such that $\sum_{i=1}^{k} s_i(A \circ B) = \text{tr}(A \circ B)C_k$. Using Lemma 3.9 and (3.19) we have

$$\sum_{i=1}^{k} s_i(A \circ B) = \text{tr}(A \circ B)C_k$$

$$= \text{tr}[A \circ (\sum_{j=1}^{n} \mu_j K_j)]C_k$$

$$= \sum_{j=1}^{n} \mu_j \text{tr}(A \circ K_j)C_k$$

$$\leq \sum_{j=1}^{n} \mu_j |\text{tr}(A \circ K_j)C_k|$$

$$\leq \sum_{j=1}^{n} \mu_j (\sum_{i=1}^{\min\{j,k\}} \alpha_i)$$

$$= \sum_{i=1}^{k} \alpha_i s_i(B).$$

This proves (3.18).   □

Lemma 3.11 *For any $A, B \in M_n$*

$$s_i(A \circ B) \leq \min\{r_i(A), c_i(A)\}s_1(B), \quad i = 1, 2, \ldots, n.$$

Proof. By Corollary 1.20

$$(A \circ B)^*(A \circ B) \leq (A^*A) \circ (B^*B).$$

Since $B^*B \leq s_1(B)^2 I$, the Schur product theorem implies

$$(A^*A) \circ (B^*B) \leq (A^*A) \circ (s_1(B)^2 I).$$

Thus

$$(A \circ B)^*(A \circ B) \leq (A^*A) \circ (s_1(B)^2 I). \tag{3.20}$$

Note that $0 \leq X \leq Y \Rightarrow s_i(X) \leq s_i(Y)$, $i = 1, 2, \ldots$. By the definition of $c_i(A)$, (3.20) implies

$$s_i(A \circ B) \leq c_i(A)s_1(B). \tag{3.21}$$

Now replace A and B in (3.21) by their adjoints A^*, B^* to get

$$s_i(A \circ B) \leq r_i(A)s_1(B).$$

This completes the proof. $\qquad \square$

Proof of Theorem 3.6. Set $\alpha_i = \min\{r_i(A), c_i(A)\}$. Lemma 3.11 gives

$$s(A \circ B) \prec_w \{\alpha_i s_1(B)\}.$$

Then applying Lemma 3.10 shows (3.16). This completes the proof. $\qquad \square$

A norm $\| \cdot \|$ on M_n is called *unitarily invariant* if

$$\|UAV\| = \|A\|$$

for all $A \in M_n$ and all unitary $U, V \in M_n$. The Fan dominance principle (Lemma 4.2 in the next chapter) says that for $A, B \in M_n$, $\|A\| \leq \|B\|$ for all unitarily invariant norms if and only if $s(A) \prec_w s(B)$.

Let us consider an application of Theorem 3.6. By a diagonal of a matrix $A = (a_{ij})$ we mean the main diagonal, or a superdiagonal, or a subdiagonal, that is, a set of all the entries a_{ij} with $i - j$ being a fixed number. Let Φ_k be an operation on M_n which keeps any but fixed k diagonals of a matrix and changes all other entries to zero, $1 \leq k < 2n - 1$. Denote by $E \in M_n$ the matrix with all entries equal to 1. Then for any $A \in M_n$, $\Phi_k(A) = \Phi_k(E) \circ A$. Applying Theorem 3.6 yields

$$\|\Phi_k(A)\| \leq \sqrt{k}\|A\|$$

for all unitarily invariant norms. In particular, if $T(A) \in M_n$ is the tridiagonal part of $A \in M_n$, then

$$\|T(A)\| \le \sqrt{3}\|A\|.$$

The constant $\sqrt{3}$ will be improved in Section 4.7.

Finally we pose the following two questions.

Question 3.12 *Is it true that for any given $A, B \in M_n$ there exist unitary matrices $U, V \in M_n$ such that*

$$|A \circ B| \le (U|A|U^*) \circ (V|B|V^*)?$$

The next weaker version involves only singular values.

Question 3.13 *Is it true that for any given $A, B \in M_n$ there exist unitary matrices $U, V \in M_n$ such that*

$$s_i(A \circ B) \le s_i[(U|A|U^*) \circ (V|B|V^*)], \quad i = 1, 2, \dots, n?$$

Notes and References. Theorem 3.6 is proved by X. Zhan [85]. A weaker conjecture is posed by R. A. Horn and C. R. Johnson [52, p.344].

Questions 3.12 and 3.13 are in [89].

See [52, Chapter 5] for more inequalities on the Hadamard product.

3.3 Differences of Positive Semidefinite Matrices

For positive real numbers a, b, $|a - b| \le \max\{a, b\}$. Now let us generalize this fact to matrices.

We need the following approximation characterization of singular values: For $G \in M_n$ and $1 \le j \le n$

$$s_j(G) = \min\{\|G - X\|_\infty : \operatorname{rank} X \le j - 1, X \in M_n\}. \tag{3.22}$$

Let us prove this fact. The following characterization [52, Theorem 3.1.2] is an immediate consequence of the minimax principle for eigenvalues of Hermitian matrices:

$$s_j(G) = \min_{\substack{\mathcal{K} \subset \mathbb{C}^n \\ \dim \mathcal{K} = n - j + 1}} \max_{\substack{u \in \mathcal{K} \\ \|u\| = 1}} \|Gu\|.$$

Suppose $\operatorname{rank} X \le j - 1$. Then $\dim \ker(X) \ge n - j + 1$. Choose any subspace $\mathcal{K}_0 \subset \ker(X)$ with $\dim \mathcal{K}_0 = n - j + 1$. We have

$$s_j(G) \leq \max_{\substack{u \in \mathcal{K}_0 \\ \|u\|=1}} \|Gu\| = \max_{\substack{u \in \mathcal{K}_0 \\ \|u\|=1}} \|(G-X)u\| \leq \|G-X\|_\infty.$$

On the other hand, let $G = U\text{diag}(s_1(G),\ldots,s_n(G))V$ be the singular value decomposition [52, Theorem 3.1.1] of G with U, V unitary. Then $X \equiv U\text{diag}(s_1(G),\ldots,s_{j-1}(G),0,\ldots,0)V$ satisfies $\text{rank}X \leq j-1$ and $s_j(G) = \|G-X\|_\infty$. This proves (3.22).

Denote the block diagonal matrix $\begin{bmatrix} A & 0 \\ 0 & B \end{bmatrix}$ by $A \oplus B$.

Theorem 3.14 *Let* $A, B \in M_n$ *be positive semidefinite. Then*

$$s_j(A-B) \leq s_j(A \oplus B), \quad j = 1, 2, \ldots, n. \tag{3.23}$$

First Proof. Note that $s(A \oplus B) = s(A) \cup s(B)$. It is easily verified (say, by using the spectral decompositions of A, B) that for a fixed j with $1 \leq j \leq n$ there exist $H, F \in M_n$ satisfying $0 \leq H \leq A$, $0 \leq F \leq B$, $\text{rank}H + \text{rank}F \leq j-1$ and

$$s_j(A \oplus B) = \|(A-H) \oplus (B-F)\|_\infty.$$

Thus $s_j(A \oplus B) = \max\{\|A-H\|_\infty, \|B-F\|_\infty\} \equiv \gamma$. Recall that we always denote by I the identity matrix.

Since

$$A - H \geq 0, B - F \geq 0, \text{rank}(H-F) \leq \text{rank}H + \text{rank}F \leq j-1,$$

by (3.22) we have

$$s_j(A-B) \leq \|A - B - (H-F)\|_\infty$$
$$= \|(A-H-\tfrac{\gamma}{2}I) - (B-F-\tfrac{\gamma}{2}I)\|_\infty$$
$$\leq \|(A-H) - \tfrac{\gamma}{2}I\|_\infty + \|(B-F) - \tfrac{\gamma}{2}I\|_\infty$$
$$\leq \tfrac{\gamma}{2} + \tfrac{\gamma}{2} = \gamma = s_j(A \oplus B).$$

This proves (3.23). □

Second Proof. In (3.13) of Corollary 3.3 setting

$$X = \begin{bmatrix} A^{1/2} & -B^{1/2} \\ 0 & 0 \end{bmatrix}, \quad Y = \begin{bmatrix} A^{1/2} & B^{1/2} \\ 0 & 0 \end{bmatrix}$$

yields (3.23). □

The above second proof shows that Theorem 3.14 follows from Corollary 3.3. Now let us point out that Theorem 3.14 implies Corollary 3.3, and hence they are equivalent.

Let
$$T = \begin{bmatrix} X & 0 \\ Y & 0 \end{bmatrix}, \quad U = \begin{bmatrix} I & 0 \\ 0 & -I \end{bmatrix}.$$

Then U is unitary. By Theorem 3.14,

$$2s_j \begin{bmatrix} XY^* & 0 \\ 0 & XY^* \end{bmatrix} = 2s_j \begin{bmatrix} 0 & XY^* \\ YX^* & 0 \end{bmatrix} = s_j[TT^* - U(TT^*)U^*]$$

$$\leq s_j \begin{bmatrix} TT^* & 0 \\ 0 & U(TT^*)U^* \end{bmatrix}$$

$$= s_j \begin{bmatrix} T^*T & 0 \\ 0 & T^*T \end{bmatrix}$$

$$= s_j \begin{bmatrix} (X^*X + Y^*Y) \oplus 0 & 0 \\ 0 & (X^*X + Y^*Y) \oplus 0 \end{bmatrix}.$$

Thus $2s_j(XY^*) \leq s_j(X^*X + Y^*Y)$.

In the proof of Theorem 4.25 in Section 4.3 we will see that sometimes it is convenient to use Theorem 3.14.

For positive real numbers a, b, $|a - b| \leq \max\{a, b\} \leq a + b$. How about the matrix generalization of the weaker relation $|a - b| \leq a + b$? In view of Theorem 3.14 one may formulate the following: $A, B \geq 0$ implies $s_j(A - B) \leq s_j(A + B)$, which is in general false for $j \geq 2$. Consider the example

$$A = \begin{bmatrix} 6 & -4 \\ -4 & 3 \end{bmatrix}, \quad B = \begin{bmatrix} 9 & 0 \\ 0 & 0 \end{bmatrix}.$$

$s(A - B) = \{5, 5\}$, $s(A + B) = \{16.21 \cdots, 1.78 \cdots\}$. The correct matrix analog is included in Theorem 3.16 below.

A unitarily invariant norm is usually considered as defined on M_n for all orders n by the rule

$$\|A\| = \left\| \begin{bmatrix} A & 0 \\ 0 & 0 \end{bmatrix} \right\|,$$

that is, adding or deleting zero singular values does not affect the value of the norm. In this way, Fan's dominance principle can be applied to matrices of different sizes. Since Theorem 3.14 implies $s(A - B) \prec_w s(A \oplus B)$, we have the following

Corollary 3.15 *Let $A, B \in M_n$ be positive semidefinite. Then*

$$\|A - B\| \leq \|A \oplus B\|$$

for all unitarily invariant norms.

The next result contains two weak log-majorization relations for singular values.

Theorem 3.16 *Let $A, B \in M_n$ be positive semidefinite. Then for any complex number z*

$$s(A - |z|B) \prec_{wlog} s(A + zB) \prec_{wlog} s(A + |z|B). \qquad (3.24)$$

Proof. We first prove the following determinant inequality for positive semidefinite matrices P, Q of the same order:

$$|\det(P - |z|Q)| \leq |\det(P + zQ)|. \qquad (3.25)$$

Without loss of generality, suppose P is positive definite. Let the eigenvalues of $P^{-1}Q$ be $\lambda_1 \geq \cdots \geq \lambda_k \geq 0$. Then

$$\begin{aligned}
|\det(P - |z|Q)| &= |\det P \cdot \det(I - |z|P^{-1}Q)| \\
&= \det P \prod_i |1 - |z|\lambda_i| \\
&\leq \det P \prod_i |1 + z\lambda_i| \\
&= |\det P \cdot \det(I + zP^{-1}Q)| \\
&= |\det(P + zQ)|.
\end{aligned}$$

This shows (3.25).

Since $A - |z|B$ is Hermitian, for $1 \leq k \leq n$ there exists an $n \times k$ matrix U such that $U^*U = I$ and

$$\prod_{j=1}^{k} s_j(A - |z|B) = |\det[U^*(A - |z|B)U]|.$$

Using (3.25) and the fact that for any $G \in M_n$, $s_j(U^*GU) \leq s_j(G)$, $j = 1, \ldots, k$, we have

$$\begin{aligned}
\prod_{j=1}^{k} s_j(A - |z|B) &= |\det[U^*(A - |z|B)U]| \\
&= |\det(U^*AU - |z|U^*BU)| \\
&\leq |\det(U^*AU + zU^*BU)| \\
&= \prod_{j=1}^{k} s_j[U^*(A + zB)U] \\
&\leq \prod_{j=1}^{k} s_j(A + zB).
\end{aligned}$$

In the third equality above we have used the fact that for any $F \in M_k$, $|\det F| = \prod_{j=1}^{k} s_j(F)$. This proves the first part of (3.24).

Recall [17, p.268] that a continuous complex-valued function f on M_n is said to be a Lieb function if it satisfies the following two conditions:
(i) $0 \leq f(A) \leq f(B)$ if $0 \leq A \leq B$;
(ii) $|f(A^*B)|^2 \leq f(A^*A)f(B^*B)$ for all A, B.

It is known [55, Theorem 6] that if $N, R \in M_n$ are normal, then for every Lieb function f on M_n

$$|f(N + R)| \leq f(|N| + |R|). \tag{3.26}$$

It is easy to verify (see [17, p.269]) that $f(G) \equiv \prod_{j=1}^{k} s_j(G)$ is a Lieb function. Applying (3.26) to this f with $N = A$, $R = zB$ yields

$$s(A + zB) \prec_{wlog} s(A + |z|B).$$

This completes the proof. $\square$

Since weak log-majorization is stronger than weak majorization, Theorem 3.16 implies the following

Corollary 3.17 *Let $A, B \in M_n$ be positive semidefinite. Then for any complex number z and any unitarily invariant norm,*

$$\|A - |z|B\| \leq \|A + zB\| \leq \|A + |z|B\|.$$

Notes and References. Corollaries 3.15 and 3.17 are due to R. Bhatia and F. Kittaneh [22, 23]. See also [17, p.280] for Corollary 3.15. Theorems 3.14 and 3.16 are proved in X. Zhan [90]. The formula (3.22) can be found in [40, p.29].

3.4 Matrix Cartesian Decompositions

In this section $i \equiv \sqrt{-1}$. Every matrix T can be written uniquely as $T = A + iB$ with A, B Hermitian:

$$A = \frac{T + T^*}{2}, \quad B = \frac{T - T^*}{2i}. \tag{3.27}$$

This is the matrix Cartesian decomposition. A and B are called the real and imaginary parts of T. We will study the relations between the eigenvalues of A, B and the singular values of T.

We need the following two results [17, p.71 and p.24] on eigenvalues of Hermitian matrices. Note that we always denote the eigenvalues of a Hermitian matrix A in decreasing order: $\lambda_1(A) \geq \cdots \geq \lambda_n(A)$ and write $\lambda(A) \equiv (\lambda_1(A), \ldots, \lambda_n(A))$.

Lemma 3.18 (Lidskii) *Let G, $H \in M_n$ be Hermitian. Then*

$$\{\lambda_j(G) + \lambda_{n-j+1}(H)\} \prec \lambda(G + H) \prec \{\lambda_j(G) + \lambda_j(H)\}.$$

Lemma 3.19 (The Minimum Principle) *Let $A \in M_n$ be Hermitian. Then for every $1 \leq k \leq n$*

$$\sum_{j=n-k+1}^{n} \lambda_j(A) = \min\{\operatorname{tr} U^* A U : U^* U = I, \ U \in M_{n,k}\}$$

where $M_{n,k}$ is the space of $n \times k$ matrices.

Throughout this section we consider the Cartesian decomposition $T = A + iB \in M_n$, and denote by $s_1 \geq \cdots \geq s_n$ the singular values of T, by α_j and β_j the eigenvalues of A and B respectively ordered in such a way that $|\alpha_1| \geq \cdots \geq |\alpha_n|$ and $|\beta_1| \geq \cdots \geq |\beta_n|$.

Theorem 3.20 *The following majorization relations hold:*

$$\{|\alpha_j + i\beta_{n-j+1}|^2\} \prec \{s_j^2\}, \tag{3.28}$$

$$\{(s_j^2 + s_{n-j+1}^2)/2\} \prec \{|\alpha_j + i\beta_j|^2\}. \tag{3.29}$$

Proof. By Lemma 3.18 with $G = A^2$ and $H = B^2$ we have

$$\{|\alpha_j + i\beta_{n-j+1}|^2\} \prec \{s_j(A^2 + B^2)\} \prec \{|\alpha_j + i\beta_j|^2\}. \tag{3.30}$$

Next note that

$$A^2 + B^2 = (T^*T + TT^*)/2$$

and

$$s_j(T^*T) = s_j(TT^*) = s_j^2.$$

Applying Lemma 3.18 again with $G = T^*T/2$, $H = TT^*/2$ gives

$$\{(s_j^2 + s_{n-j+1}^2)/2\} \prec \{s_j(A^2 + B^2)\} \prec \{s_j^2\}. \tag{3.31}$$

Combining (3.30) and (3.31) we get (3.28) and (3.29). □

An important class of unitarily invariant norms are *Schatten p-norms*. They are the l_p norms of the singular values:

$$\|A\|_p \equiv \left(\sum_{j=1}^{n} s_j(A)^p \right)^{1/p}, \quad p \geq 1, \ A \in M_n.$$

Note that the case $p = \infty$ of this notation coincides with what we have used: $\|A\|_\infty = s_1(A)$ is just the spectral norm. $\|\cdot\|_2$ is the Frobenius (or Hilbert-Schmidt) norm: $\|A\|_2 = (\operatorname{tr} AA^*)^{1/2} = (\sum_{j,k} |a_{jk}|^2)^{1/2}$ for $A = (a_{jk})$ and $\|\cdot\|_1$ is called the trace norm.

Theorem 3.21 *Let $T = A + iB$ with A, B Hermitian. Then*

$$2^{2/p-1}(\|A\|_p^2 + \|B\|_p^2) \le \|T\|_p^2 \le 2^{1-2/p}(\|A\|_p^2 + \|B\|_p^2) \tag{3.32}$$

for $2 \le p \le \infty$ and

$$2^{2/p-1}(\|A\|_p^2 + \|B\|_p^2) \ge \|T\|_p^2 \ge 2^{1-2/p}(\|A\|_p^2 + \|B\|_p^2) \tag{3.33}$$

for $1 \le p \le 2$.

Proof. When $p \ge 2$, $f(t) = t^{p/2}$ on $[0, \infty)$ is convex and when $1 \le p \le 2$, $g(t) = -t^{p/2}$ on $[0, \infty)$ is convex. Using Theorem 2.2 we get from (3.29)

$$\{2^{-p/2}(s_j^2 + s_{n-j+1}^2)^{p/2}\} \prec_w \{|\alpha_j + i\beta_j|^p\} \quad \text{for} \quad p \ge 2,$$

$$\{-2^{-p/2}(s_j^2 + s_{n-j+1}^2)^{p/2}\} \prec_w \{-|\alpha_j + i\beta_j|^p\} \quad \text{for} \quad 1 \le p \le 2.$$

In particular, we get

$$\sum_{j=1}^n (s_j^2 + s_{n-j+1}^2)^{p/2} \le 2^{p/2} \sum_{j=1}^n |\alpha_j + i\beta_j|^p \quad \text{for} \quad p \ge 2, \tag{3.34}$$

$$\sum_{j=1}^n (s_j^2 + s_{n-j+1}^2)^{p/2} \ge 2^{p/2} \sum_{j=1}^n |\alpha_j + i\beta_j|^p \quad \text{for } 1 \le p \le 2. \tag{3.35}$$

Since for fixed nonnegative real numbers a, b, the function $t \mapsto (a^t + b^t)^{1/t}$ is decreasing on $(0, \infty)$,

$$s_j^p + s_{n-j+1}^p \le (s_j^2 + s_{n-j+1}^2)^{p/2} \quad \text{for} \quad p \ge 2$$

and this inequality is reversed when $1 \le p \le 2$. Hence from (3.34) and (3.35) we obtain

$$\sum_{j=1}^n s_j^p \le 2^{p/2-1} \sum_{j=1}^n |\alpha_j + i\beta_j|^p \quad \text{for} \quad p \ge 2, \tag{3.36}$$

$$\sum_{j=1}^n s_j^p \ge 2^{p/2-1} \sum_{j=1}^n |\alpha_j + i\beta_j|^p \quad \text{for } 1 \le p \le 2. \tag{3.37}$$

Finally applying the Minkowski inequality

$$\left(\sum_j (x_j + y_j)^r\right)^{1/r} \le \left(\sum_j x_j^r\right)^{1/r} + \left(\sum_j y_j^r\right)^{1/r}, \quad r \ge 1,$$

$$\left(\sum_j (x_j + y_j)^r\right)^{1/r} \geq \left(\sum_j x_j^r\right)^{1/r} + \left(\sum_j y_j^r\right)^{1/r}, \quad 0 < r \leq 1$$

for nonnegative sequences $\{x_j\}, \{y_j\}$ to the above two inequalities yields the right-hand sides of (3.32) and (3.33). The left-hand sides of these two inequalities can be deduced from the majorization (3.28) in a similar way. □

From the majorization relations in (3.31), using a similar argument as above we can derive the following two inequalities:

$$\|(A^2 + B^2)^{1/2}\|_p \leq \|T\|_p \leq 2^{1/2 - 1/p}\|(A^2 + B^2)^{1/2}\|_p \qquad (3.38)$$

for $2 \leq p \leq \infty$ and

$$\|(A^2 + B^2)^{1/2}\|_p \geq \|T\|_p \geq 2^{1/2 - 1/p}\|(A^2 + B^2)^{1/2}\|_p \qquad (3.39)$$

for $1 \leq p \leq 2$.
The example

$$T = \begin{bmatrix} 0 & 0 \\ 2 & 0 \end{bmatrix}, \quad A = \begin{bmatrix} 0 & 1 \\ 1 & 0 \end{bmatrix}, \quad B = \begin{bmatrix} 0 & i \\ -i & 0 \end{bmatrix}$$

shows that the second inequalities in (3.32), (3.33), (3.38) and (3.39) are sharp, while the example

$$A = \begin{bmatrix} 1 & 0 \\ 0 & 0 \end{bmatrix}, \quad B = \begin{bmatrix} 0 & 0 \\ 0 & 1 \end{bmatrix}$$

shows that the first inequalities in both (3.32) and (3.33) are sharp. Note that in this latter example, A and B are positive semidefinite, so even in this special case, the first inequalities in both (3.32) and (3.33) are also sharp. But as we will see soon, when A or (and) B is positive semidefinite, the second inequalities in (3.32) and (3.33) can be improved. The first inequalities in (3.38) and (3.39) are obviously sharp, since they are equalities when the matrices are scalars.

Theorem 3.22 *Let $T = A + iB \in M_n$ where A and B are Hermitian with respective eigenvalues α_j and β_j ordered so that $|\alpha_1| \geq \cdots \geq |\alpha_n|$ and $|\beta_1| \geq \cdots \geq |\beta_n|$. Then*

$$|\det T| \leq \prod_{j=1}^n |\alpha_j + i\beta_{n-j+1}|.$$

Proof. Since the set of invertible matrices is dense in M_n, we may assume that T is invertible. Then every $s_j > 0$ and (3.28) implies $|\alpha_j + i\beta_{n-j+1}| > 0$ for each j. Applying the convex function $f(t) = -\frac{1}{2}\log t$ on $(0, \infty)$ to the majorization (3.28) completes the proof. □

Now we consider the case when one of the real and imaginary parts, say, the real part A, is positive semidefinite. For the case when $B \geq 0$, by considering $-iT$ we return to the former case and all the results are similar.

We take the same argument line as above: First establish a majorization relation, and then apply majorization principles to obtain inequalities for Schatten p-norms. The trace norm case will be treated separately.

Continue to use the notation s_j, α_j, β_j. From $\{\beta_j\}$ we define another nonnegative n-tuple $\{\gamma_j\}$ as follows. If n is an even number and $n = 2m$, then

$$\gamma_j^2 \equiv \begin{cases} \beta_{2j-1}^2 + \beta_{2j}^2 & \text{for} \quad 1 \leq j \leq m, \\ 0 & \text{for} \quad m+1 \leq j \leq n. \end{cases}$$

If n is an odd number and $n = 2m + 1$, then

$$\gamma_j^2 \equiv \begin{cases} \beta_{2j-1}^2 + \beta_{2j}^2 & \text{for} \quad 1 \leq j \leq m, \\ \beta_n^2 & \text{for} \quad j = m+1, \\ 0 & \text{for} \quad m+2 \leq j \leq n. \end{cases}$$

Theorem 3.23 Let $T = A + iB \in M_n$ with A positive semidefinite and B Hermitian. Then we have the majorization

$$\{s_j^2\} \prec \{\alpha_j^2 + \gamma_j^2\}. \tag{3.40}$$

Note that the relation (3.40) is equivalent to the following three statements together:

$$\sum_{j=1}^{k} s_j^2 \leq \sum_{j=1}^{k} \alpha_j^2 + \sum_{j=1}^{2k} \beta_j^2 \quad \text{for} \quad 1 \leq k < n/2, \tag{3.41}$$

$$\sum_{j=1}^{k} s_j^2 \leq \sum_{j=1}^{k} \alpha_j^2 + \sum_{j=1}^{n} \beta_j^2 \quad \text{for} \quad n/2 \leq k \leq n, \tag{3.42}$$

$$\sum_{j=1}^{n} s_j^2 = \sum_{j=1}^{n} \alpha_j^2 + \sum_{j=1}^{n} \beta_j^2. \tag{3.43}$$

Of these, the equation (3.43) is a restatement of

$$\|T\|_2^2 = \|A\|_2^2 + \|B\|_2^2 \tag{3.44}$$

which is easy to verify.

For the proof we need the following lemma.

Lemma 3.24 Let k, n be positive integers with $1 \leq k < n/2$. Let $\mu_1 \geq \mu_2 \geq \cdots \geq \mu_n$ and $\lambda_1 \geq \lambda_2 \geq \cdots \geq \lambda_{n-k}$ be real numbers satisfying the interlacing inequalities

$$\mu_j \geq \lambda_j \geq \mu_{j+k} \quad \text{for} \quad j = 1, 2, \ldots, n - k.$$

Then we can choose $n - 2k$ *distinct indices* $j_1, \ldots, j_{n-2k}$ *out of the set* $\{1, \ldots, n - k\}$ *so that*

$$|\lambda_{j_s}| \geq |\mu_{k+s}| \quad \text{for} \quad s = 1, 2, \ldots, n - 2k.$$

Proof. Three different cases can arise. We list them and the corresponding choices in each case.

(i) If $\mu_{n-k} \geq 0$, choose

$$\{j_1, \ldots, j_{n-2k}\} = \{1, 2, \ldots, n - 2k\}.$$

(ii) If for some r with $n - k > r \geq k + 1$, we have $\mu_r \geq 0 > \mu_{r+1}$, choose

$$\{j_1, \ldots, j_{n-2k}\} = \{1, \ldots, r - k, r + 1, \ldots, n - k\}.$$

(iii) If $0 > \mu_{k+1}$, choose

$$\{j_1, \ldots, j_{n-2k}\} = \{k + 1, \ldots, n - k\}.$$

In each case the assertion of the lemma is readily verified.   □

Proof of Theorem 3.23. Let $k < n/2$. Because of (3.43), the inequality (3.41) is equivalent to

$$\sum_{j=k+1}^{n} s_j^2 \geq \sum_{j=k+1}^{n} \alpha_j^2 + \sum_{j=2k+1}^{n} \beta_j^2. \tag{3.45}$$

By Lemma 3.19 there exists an $n \times (n - k)$ matrix U with $U^*U = I$ such that

$$\sum_{j=k+1}^{n} s_j^2 = \operatorname{tr} U^*T^*TU. \tag{3.46}$$

Since $UU^* \leq I$,

$$\operatorname{tr} U^*T^*TU \geq \operatorname{tr} U^*T^*U \cdot U^*TU.$$

From this, we obtain using the relation (3.44) (with U^*TU in place of T)

$$\operatorname{tr} U^*T^*TU \geq \operatorname{tr} (U^*AU)^2 + \operatorname{tr} (U^*BU)^2. \tag{3.47}$$

Let V be an $n \times k$ matrix such that $W \equiv (U, V)$ is unitary. Then U^*AU and U^*BU are principal submatrices of W^*AW and W^*BW respectively. Note also that $\lambda(W^*AW) = \lambda(A)$ and $\lambda(W^*BW) = \lambda(B)$. Hence by Cauchy's interlacing theorem for eigenvalues of Hermitian matrices [17, Corollary III.1.5] we have

$$\lambda_j(U^*AU) \geq \lambda_{j+k}(A), \quad 1 \leq j \leq n-k,$$

and

$$\lambda_j(B) \geq \lambda_j(U^*BU) \geq \lambda_{j+k}(B), \quad 1 \leq j \leq n-k.$$

The first of these inequalities shows that

$$\operatorname{tr}(U^*AU)^2 = \sum_{j=1}^{n-k}[\lambda_j(U^*AU)]^2 \geq \sum_{j=k+1}^{n} \alpha_j^2. \tag{3.48}$$

Applying Lemma 3.24 to the second inequality, we deduce that there exist $n-2k$ distinct indices $j_1, \ldots, j_{n-2k}$ such that

$$|\lambda_{j_s}(U^*BU)| \geq |\lambda_{k+s}(B)|, \quad s = 1, \ldots, n-2k.$$

Therefore

$$\operatorname{tr}(U^*BU)^2 = \sum_{j=1}^{n-k}[\lambda_j(U^*BU)]^2 \geq \sum_{j=2k+1}^{n} \beta_j^2. \tag{3.49}$$

Combining (3.46)-(3.49) we obtain the desired inequality (3.45).

Now let $n/2 \leq k \leq n$. Again, because of (3.43) the inequality (3.42) is equivalent to

$$\sum_{j=k+1}^{n} s_j^2 \geq \sum_{j=k+1}^{n} \alpha_j^2.$$

The proof of this is easier. Just drop the last term in (3.47) and use (3.48). □

To apply Theorem 3.23 we need to establish a relation between the norms of the tuples $\{\gamma_j\}$ and $\{\beta_j\}$. Given an n-tuple $\delta = \{\delta_1, \ldots, \delta_n\}$ and an m-tuple $\sigma = \{\sigma_1, \ldots, \sigma_m\}$ we write $\delta \vee \sigma$ for the $(n+m)$-tuple obtained by combining them. We identify $\delta \vee 0$ with δ. We write δ^2 for the tuple $\{\delta_1^2, \ldots, \delta_n^2\}$.

Note that from the definition of $\{\gamma_j\}$ it follows that

$$\gamma^2 \vee \gamma^2 \prec 2\beta^2. \tag{3.50}$$

Lemma 3.25 *We have the inequalities*

$$\|\gamma\|_p^2 \leq 2^{1-2/p}\|\beta\|_p^2 \quad \text{for} \quad 2 \leq p \leq \infty,$$

$$\|\gamma\|_p^2 \geq 2^{1-2/p}\|\beta\|_p^2 \quad \text{for} \quad 1 \leq p \leq 2.$$

Proof. For all values of $p > 0$,

$$\|\gamma\|_p^2 = 2^{-2/p}\|\gamma \vee \gamma\|_p^2 = 2^{-2/p}\|\gamma^2 \vee \gamma^2\|_{p/2}.$$

Using familiar properties of majorization, one has from (3.50)

$$\|\gamma^2 \vee \gamma^2\|_{p/2} \le 2\|\beta^2\|_{p/2} \quad \text{for} \quad 2 \le p \le \infty,$$

and the opposite inequality for $1 \le p \le 2$. Hence for $2 \le p \le \infty$, we have

$$\|\gamma\|_p^2 \le 2^{1-2/p}\|\beta^2\|_{p/2} = 2^{1-2/p}\|\beta\|_p^2,$$

and for $1 \le p \le 2$, the inequality is reversed.   □

Theorem 3.26 *Let $T = A + iB$ with A positive semidefinite and B Hermitian. Then*

$$\|T\|_p^2 \le \|A\|_p^2 + 2^{1-2/p}\|B\|_p^2 \qquad (3.51)$$

for $2 \le p \le \infty$ and

$$\|T\|_p^2 \ge \|A\|_p^2 + 2^{1-2/p}\|B\|_p^2 \qquad (3.52)$$

for $1 \le p \le 2$.

Proof. Denote $\Gamma = \mathrm{diag}(\gamma_1, \ldots, \gamma_n)$. Let $p \ge 2$. Using the convexity of the function $f(t) = t^{p/2}$ on the positive half-line, Theorem 2.2, and the Minkowski inequality we obtain from (3.40) the inequality

$$\|T\|_p^2 \le \|A\|_p^2 + \|\Gamma\|_p^2.$$

Now use Lemma 3.25 to obtain (3.51).

All the inequalities involved in this argument are reversed for $1 \le p \le 2$. This leads to (3.52).   □

We now discuss the sharpness of the bounds (3.51) and (3.52). When $p \ge 2$, the factor $2^{1-2/p}$ occurring in (3.51) can not be replaced by any smaller number. Consider the example

$$A = \begin{bmatrix} 1 & 0 \\ 0 & 0 \end{bmatrix}, \quad B = \begin{bmatrix} 0 & t \\ t & 0 \end{bmatrix}$$

where t is a real number. We see that in this case the right-hand side of (3.51) is equal to $1 + 2t^2$ for all values of p. If for some $2 < p < \infty$, we could replace the factor $2^{1-2/p}$ in (3.51) by a smaller number, then we would have in this case

$$(1 + 2t^2 + \sqrt{1 + 4t^2})/2 = \|T\|_\infty^2 \le \|T\|_p^2 < 1 + (2 - \epsilon)t^2.$$

But the left-hand side is larger than the right-hand side for small t.

Because of symmetry considerations one would expect the inequality (3.52) to be sharp as well. But this is not the case as we will now see.

Theorem 3.27 *Let $T = A + iB$ with A positive semidefinite and B Hermitian. Then*

$$\|T\|_1^2 \ge \|A\|_1^2 + \|B\|_1^2. \qquad (3.53)$$

Proof. First consider the case when the order of matrices $n = 2$. Because of (3.44) the inequality (3.53) is equivalent to the statement

$$|\det T| \geq \det A + |\det B|. \tag{3.54}$$

By applying unitary conjugations, we may assume

$$A = \begin{bmatrix} a & c \\ c & b \end{bmatrix}, \quad B = \begin{bmatrix} s & 0 \\ 0 & t \end{bmatrix},$$

where a, b are positive and c, s, t are real numbers. The inequality (3.54) then says

$$|(ab - c^2 - st) + i(at + bs)| \geq ab - c^2 + |st|.$$

If st is negative, this is obvious. If st is positive, this inequality follows from

$$(ab - c^2 + st)^2 - (ab - c^2 - st)^2 = 4(ab - c^2)st$$
$$\leq 4abst \leq (at + bs)^2.$$

Now let the order n be arbitrary. Again, because of (3.44) the inequality (3.53) reduces to the statement

$$\sum_{j<k} s_j s_k \geq \sum_{j<k} \alpha_j \alpha_k + \sum_{j<k} |\beta_j \beta_k|,$$

where s_j, α_j and $|\beta_j|$, $1 \leq j \leq n$ are the singular values of T, A and B respectively. Let $\wedge^2 T$ denote the second antisymmetric tensor power [17] of T. Then this inequality can be restated as

$$\|\wedge^2 T\|_1 \geq \|\wedge^2 A\|_1 + \|\wedge^2 B\|_1. \tag{3.55}$$

For each pair $j < k$, let $C_{jk}(T)$ be the determinant of the 2×2 principal submatrix of T defined as $\begin{bmatrix} t_{jj} & t_{jk} \\ t_{kj} & t_{kk} \end{bmatrix}$. These are the diagonal entries of the matrix $\wedge^2 T$.

Since every unitarily invariant norm is diminished when we replace a matrix by its diagonal part (see (4.76) in Section 4.7),

$$\|\wedge^2 T\|_1 \geq \sum_{j<k} |C_{jk}(T)|.$$

By (3.54), $|C_{jk}(T)| \geq C_{jk}(A) + |C_{jk}(B)|$ for all $j < k$. By applying a unitary conjugation we may assume that B is diagonal. Then, it is easy to see from the above considerations that

$$\|\wedge^2 T\|_1 \geq \text{tr} \wedge^2 A + \text{tr}|\wedge^2 B|.$$

This proves (3.55).   □

In view of Theorem 3.27, the inequality (3.52) for $p = 1$ is not sharp. So it is natural to pose the following

Problem 3.28 *For $1 < p < 2$, determine the largest constant c_p such that*

$$\|T\|_p^2 \geq \|A\|_p^2 + c_p\|B\|_p^2$$

holds for all $T = A + iB$ with A positive semidefinite and B Hermitian. In particular, is it true that $c_p = 1$ for all $1 < p < 2$?

The majorization (3.40) can be used to derive several other inequalities. Using the convexity of the function $f(t) = -\frac{1}{2}\log t$ on the positive half-line we get from it the family of inequalities

$$\prod_{j=k}^{n} s_j \geq \prod_{j=k}^{n} (\alpha_j^2 + \gamma_j^2)^{1/2}, \quad 1 \leq k \leq n.$$

The special case $k = 1$ gives

$$|\det T| \geq \prod_{j=1}^{n} (\alpha_j^2 + \gamma_j^2)^{1/2}.$$

This sharpens a well-known inequality of Ostrowski and Taussky which says

$$|\det T| \geq \det A = \prod_{j=1}^{n} \alpha_j$$

whenever $T = A + iB$ with A positive semidefinite and B Hermitian.

Next we consider the still more special case when both the real and the imaginary parts are positive semidefinite. Reasonably the results for this case have the most pleasant form.

Theorem 3.29 *Let $T = A + iB \in M_n$ where A and B are positive semidefinite. Then*

$$\{s_j^2\} \prec \{\alpha_j^2 + \beta_j^2\}. \tag{3.56}$$

Proof. By (3.43), to prove (3.56) it suffices to show

$$\sum_{j=n-k+1}^{n} s_j^2 \geq \sum_{j=n-k+1}^{n} (\alpha_j^2 + \beta_j^2), \quad 1 \leq k \leq n. \tag{3.57}$$

The left-hand side of (3.57) has an extremal representation (see Lemma 3.19)

$$\sum_{j=n-k+1}^{n} s_j^2 = \min\{\operatorname{tr} U^*T^*TU : U \in M_{n,k}, U^*U = I\}. \tag{3.58}$$

From the condition $U^*U = I$ we get $I \geq UU^*$. This implies

$$
\begin{aligned}
U^*T^*TU = U^*T^* \cdot I \cdot TU &\geq U^*T^* \cdot UU^* \cdot TU \\
&= U^*(A - iB)U \cdot U^*(A + iB)U \\
&= (U^*AU)^2 + (U^*BU)^2 + i[U^*AU, U^*BU]
\end{aligned}
$$

where $[X, Y]$ stands for the commutator $XY - YX$. It follows that

$$
\operatorname{tr} U^*T^*TU \geq \operatorname{tr} (U^*AU)^2 + \operatorname{tr} (U^*BU)^2. \tag{3.59}
$$

The operator U^*AU is a compression of A to a k-dimensional subspace. Hence, by Cauchy's interlacing theorem [17, Corollary III.1.5] we have

$$
\lambda_j(U^*AU) \geq \alpha_{j+n-k}, \quad 1 \leq j \leq k.
$$

By the same argument

$$
\lambda_j(U^*BU) \geq \beta_{j+n-k}, \quad 1 \leq j \leq k.
$$

So, from (3.59) we have

$$
\operatorname{tr} U^*T^*TU \geq \sum_{j=n-k+1}^{n} (\alpha_j^2 + \beta_j^2). \tag{3.60}
$$

The inequality (3.57) follows from (3.58) and (3.60). □

As before, the next two theorems follow as corollaries.

Theorem 3.30 *Let $T = A + iB$ where A and B are positive semidefinite. Then*

$$
\|T\|_p^2 \leq \|A\|_p^2 + \|B\|_p^2 \quad \text{for} \quad 2 \leq p \leq \infty, \tag{3.61}
$$

$$
\|T\|_p^2 \geq \|A\|_p^2 + \|B\|_p^2 \quad \text{for} \quad 1 \leq p \leq 2. \tag{3.62}
$$

Proof. Using Theorem 2.2, apply the convex functions $f(t) = t^{p/2}$ $(p \geq 2)$, $g(t) = -t^{p/2}$ $(1 \leq p \leq 2)$ to the majorization (3.56), and then use the Minkowski inequality. □

The inequalities (3.61) and (3.62) are obviously sharp.

Theorem 3.31 *Let $T = A + iB$ where A and B are positive semidefinite. Then*

$$
\prod_{j=k}^{n} s_j \geq \prod_{j=k}^{n} |\alpha_j + i\beta_j|, \quad 1 \leq k \leq n. \tag{3.63}
$$

Proof. Using Theorem 2.2, apply the convex function $f(t) = -\frac{1}{2}\log t$ on the positive half-line to the majorization (3.56). □

Note that the case $k = 1$ of (3.63) can be written as

$$|\det T| \geq \prod_{j=1}^{n} |\alpha_j + i\beta_j|. \tag{3.64}$$

Looking at the inequalities in Theorems 3.21, 3.26 and 3.30, we see a complete picture for the three cases, from the general to the special.

Notes and References. Theorem 3.20 is due to T. Ando and R. Bhatia [9]. Theorem 3.21 and the inequalities (3.38), (3.39) are proved by R. Bhatia and F. Kittaneh [25] using a different approach. Theorem 3.22 is due to J. F. Queiró and A. L. Duarte [80]; the proof given here is in [9]. The inequality (3.64) is due to N. Bebiano.

All the other results in this section are proved by R. Bhatia and X. Zhan in the two papers [27] and [28].

3.5 Singular Values and Matrix Entries

Let $\lambda_1, \ldots, \lambda_n$ be the eigenvalues of a matrix $A = (a_{ij}) \in M_n$. Then Schur's inequality says that

$$\sum_{j=1}^{n} |\lambda_j|^2 \leq \sum_{i,j=1}^{n} |a_{ij}|^2 = \|A\|_2^2$$

and the equality is attained if and only if A is normal. This is a relation between eigenvalues and matrix entries.

What can be said about the relationship between singular values and matrix entries? Two obvious relations are the following: The modulus of each entry is not greater than the largest singular value (the spectral norm), i.e.,

$$\max_{i,j} |a_{ij}| \leq s_1(A) \quad \text{and} \quad \sum_{i,j=1}^{n} |a_{ij}|^2 = \sum_{j=1}^{n} s_j(A)^2.$$

Now let us prove more.

Theorem 3.32 *Let $s_1, s_2, \ldots, s_n$ be the singular values of a matrix $A = (a_{ij}) \in M_n$. Then*

$$\sum_{j=1}^{n} s_j^p \leq \sum_{i,j=1}^{n} |a_{ij}|^p \quad \text{for} \quad 0 < p \leq 2, \tag{3.65}$$

$$\sum_{j=1}^{n} s_j^p \geq \sum_{i,j=1}^{n} |a_{ij}|^p \quad \text{for} \quad p \geq 2. \tag{3.66}$$

Proof. Let $c_j = (\sum_{i=1}^n |a_{ij}|^2)^{1/2}$ be the Euclidean length of the jth column of A. Then $c_1^2, \ldots, c_n^2$ and $s_1^2, \ldots, s_n^2$ are respectively the diagonal entries and eigenvalues of A^*A. By Schur's theorem (Theorem 2.1) we have

$$\{c_j^2\} \prec \{s_j^2\}. \tag{3.67}$$

First consider the case $p \geq 2$. Since the function $f(t) = t^{p/2}$ is convex on $[0\,\infty)$, applying Theorem 2.2 with $f(t)$ to the majorization (3.67) yields

$$\{c_j^p\} \prec_w \{s_j^p\}.$$

In particular,

$$\sum_{j=1}^n c_j^p \leq \sum_{j=1}^n s_j^p. \tag{3.68}$$

On the other hand we have

$$\sum_{i=1}^n |a_{ij}|^p \leq (\sum_{i=1}^n |a_{ij}|^2)^{p/2} = c_j^p, \quad j = 1, \ldots, n$$

and hence

$$\sum_{i,j=1}^n |a_{ij}|^p \leq \sum_{j=1}^n c_j^p. \tag{3.69}$$

Combining (3.69) and (3.68) gives (3.66).

When $0 < p \leq 2$, by considering the convex function $g(t) = -t^{p/2}$, on $[0\,\infty)$, all the above inequalities are reversed. This proves (3.65). □

Note that taking the $1/p$-th power of both sides of (3.66) and letting $p \to \infty$ we recapture the fact that $\max_{i,j} |a_{ij}| \leq \|A\|_\infty$.

Let $\lambda_1, \ldots, \lambda_n$ be the eigenvalues of a matrix A. By Weyl's theorem (Theorem 2.4) we have

$$\{|\lambda_j|^p\} \prec_{log} \{s_j^p\} \quad \text{for any } p > 0.$$

Since weak log-majorization implies majorization (Theorem 2.7), we get

$$\{|\lambda_j|^p\} \prec_w \{s_j^p\} \quad \text{for any } p > 0.$$

In particular,

$$\sum_{j=1}^n |\lambda_j|^p \leq \sum_{j=1}^n s_j^p \quad \text{for any } p > 0. \tag{3.70}$$

Combining (3.70) and (3.65) we get the next

Corollary 3.33 *Let $\lambda_1, \ldots, \lambda_n$ be the eigenvalues of a matrix $A = (a_{ij}) \in M_n$. Then*

$$\sum_{j=1}^{n} |\lambda_j|^p \leq \sum_{i,j=1}^{n} |a_{ij}|^p \quad \text{for} \quad 0 < p \leq 2. \tag{3.71}$$

Observe that Schur's inequality mentioned at the beginning of this section corresponds to the case $p = 2$ of (3.71).

Next let us derive two entrywise matrix inequalities, one of which is an application of Theorem 3.32.

For two n-square real matrices A, B, we write $A \leq_e B$ to mean that $B - A$ is (entrywise) nonnegative. Denote $\mathcal{N} \equiv \{1, 2, \ldots, n\}$. For $\alpha \subseteq \mathcal{N}$, α^c denotes the complement $\mathcal{N}\backslash\alpha$ and $|\alpha|$ the cardinality of α. Given $\alpha, \beta \subseteq \mathcal{N}$ and $A = (a_{ij}) \in M_n$, we write $A(\alpha, \beta) \equiv \sum_{i \in \alpha, j \in \beta} a_{ij}$.

A proof of the following lemma can be found in [5, p.192].

Lemma 3.34 *Assume that the n-square nonnegative matrices B and C satisfy $C \leq_e B$. Then there exists a doubly stochastic matrix A such that*

$$C \leq_e A \leq_e B$$

if and only if

$$B(\alpha, \beta) \geq C(\alpha^c, \beta^c) + |\alpha| + |\beta| - n \text{ for all } \alpha, \beta \subseteq \mathcal{N}.$$

A nonnegative matrix Z is called *doubly superstochastic* if there is a doubly stochastic matrix A such that $A \leq_e Z$; correspondingly a nonnegative matrix W is called *doubly substochastic* if there is a doubly stochastic matrix A such that $W \leq_e A$. The following corollary is an immediate consequence of Lemma 3.34.

Corollary 3.35 *Let Z, W be n-square nonnegative matrices. Then*
  (i) Z is doubly superstochastic if and only if

$$Z(\alpha, \beta) \geq |\alpha| + |\beta| - n \text{ for all } \alpha, \beta \subseteq \mathcal{N};$$

  (ii) W is doubly substochastic if and only if all row sums and column sums of W are less than or equal to 1.

For $A = (a_{ij}) \in M_n$ and a real number $p > 0$, we denote $A^{|\circ|p} \equiv (|a_{ij}|^p) \in M_n$. The following entrywise inequalities involve the smallest and the largest singular values.

Theorem 3.36 *Let $A \in M_n$ and let p, q be real numbers with $0 < p \leq 2$ and $q \geq 2$. Then there exist two doubly stochastic matrices $B, C \in M_n$ such that*

$$s_n(A)^p B \leq_e A^{|\circ|p} \tag{3.72}$$

and

$$A^{|\circ|q} \le_e s_1(A)^q C. \tag{3.73}$$

Proof. In view of Corollary 3.35(i), to prove (3.72) it suffices to show

$$A^{|\circ|p}(\alpha, \beta) \ge (|\alpha| + |\beta| - n)s_n(A)^p \tag{3.74}$$

for all , $\alpha, \beta \subseteq \mathcal{N}$.

For an $r \times t$ matrix X, we define its singular values to be the eigenvalues of $(X^*X)^{1/2}$ or $(XX^*)^{1/2}$ according as $r \ge t$ or $r < t$. This is just to avoid obvious zero singular values. It is known [52, Corollary 3.1.3] that if X is an $r \times t$ submatrix of an n-square matrix Y and $r + t - n \ge 1$ then

$$s_j(X) \ge s_{j+2n-r-t}(Y) \quad \text{for} \quad j \le r + t - n.$$

In particular

$$s_{r+t-n}(X) \ge s_n(Y). \tag{3.75}$$

Since nonsquare matrices can be augmented to square ones by adding zero rows or columns, Theorem 3.32 is also true for rectangular matrices. For instance, given an $r \times t$ matrix $X = (x_{ij})$, the inequality (3.65) has the following form:

$$\sum_{i=1}^{\min(r,t)} s_i(X)^p \le \sum_{i=1}^{r}\sum_{j=1}^{t} |x_{ij}|^p \quad \text{for} \quad 0 < p \le 2. \tag{3.76}$$

Now we prove (3.74). If $|\alpha| + |\beta| - n \le 0$, (3.74) holds trivially. Assume $|\alpha| + |\beta| - n \ge 1$. Denote by $A[\alpha|\beta]$ the submatrix of A with rows and columns indexed respectively by α and β. Then from (3.75) we have

$$s_1(A[\alpha|\beta]) \ge \cdots \ge s_{|\alpha|+|\beta|-n}(A[\alpha|\beta]) \ge s_n(A). \tag{3.77}$$

Let $A = (a_{ij})$. Using (3.76) and (3.77) we obtain

$$A^{|\circ|p}(\alpha, \beta) = \sum_{i\in\alpha, j\in\beta} |a_{ij}|^p$$

$$\ge \sum_{i=1}^{\min(|\alpha|, |\beta|)} s_i(A[\alpha|\beta])^p$$

$$\ge (|\alpha| + |\beta| - n)s_n(A)^p.$$

This proves (3.74) and hence (3.72).

Denote by $\nu(X)$ the largest of the row and column sums of a nonnegative matrix X. By Corollary 3.35(ii), the following inequality

$$\nu(A^{|\circ|q}) \le s_1(A)^q \tag{3.78}$$

will imply the existence of a doubly stochastic matrix C satisfying (3.73). But (3.78) is equivalent to

$$[\nu(A^{[\circ|q]})]^{1/q} \leq s_1(A)$$

which is obvious since for $q \geq 2$,

$$[\nu(A^{[\circ|q]})]^{1/q} \leq [\nu(A^{[\circ|2]})]^{1/2} \leq s_1(A).$$

This proves (3.73).   □

We remark that in (3.72) p can not be taken as $p > 2$ and in (3.73) q can not be taken as $q < 2$. To see this just consider any unitary matrix that is not a generalized permutation matrix and compare the row or column sums of both sides in (3.72) and (3.73).

Notes and References. Theorem 3.32 is based on K. D. Ikramov [56] where the case $1 \leq p \leq 2$ is treated. The case $p = q = 2$ of Theorem 3.36 is due to L. Elsner and S. Friedland [34]; the theorem itself seems new.

4. Norm Inequalities

Recent research on norm inequalities is very active. Such inequalities are not only of theoretical interest but also of practical importance.

On the one hand, all norms on a finite-dimensional space are equivalent in the sense that for any two norms $\|\cdot\|_\alpha$ and $\|\cdot\|_\beta$ there exist positive constants c_1 and c_2 such that

$$c_1\|x\|_\alpha \le \|x\|_\beta \le c_2\|x\|_\alpha \quad \text{for all } x.$$

Thus all norms on a finite-dimensional vector space generate the same topology. So for topological problems, e.g., the convergence of a sequence of matrices or vectors which is of primary significance for iterative methods, different choices of norms have equal effect. On the other hand, in applications (numerical analysis, say) to solve a particular problem it may be more convenient to use one norm than the other. For example, the spectral norm has good geometric properties but it is difficult to compute its accurate value, while the Frobenius norm is easy to compute but might not be suitable for describing the problem. Therefore we need various kinds of norms.

Recall that a norm $\|\cdot\|$ on M_n is called *unitarily invariant* if $\|UAV\| = \|A\|$ for all A and all unitary U, V. The singular value decomposition theorem [52, Theorem 3.1.1] asserts that for any given $A \in M_n$, there are unitary matrices U, V such that $A = U\mathrm{diag}(s_1(A), \ldots, s_n(A))V$. Thus unitarily invariant norms are functions of singular values. von Neumann (see [17] and [52]) proved that these are *symmetric gauge functions,* i.e., those norms $\varPhi$ on $\mathbb{R}^n$ satisfying

(i) $\varPhi(Px) = \varPhi(x)$ for all permutation matrices P and $x \in \mathbb{R}^n$,
(ii) $\varPhi(\epsilon_1 x_1, \ldots, \epsilon_n x_n) = \varPhi(x_1, \ldots, x_n)$ if $\epsilon_j = \pm 1$.

In other words, we have a one to one correspondence between unitarily invariant norms $\|\cdot\|_\varPhi$ and symmetric gauge functions $\varPhi$ and they are related in the following way:

$$\|A\|_\varPhi = \varPhi(s_1(A), \ldots, s_n(A)) \quad \text{for all } A \in M_n.$$

In the preceding chapter we have talked about a class of unitarily invariant norms: the Schatten p-norms. Another very important class of unitarily invariant norms are the Fan k-norms defined as

$$\|A\|_{(k)} = \sum_{j=1}^{k} s_j(A), \quad 1 \le k \le n.$$

Note that in this notation, $\|\cdot\|_{(1)} = \|\cdot\|_\infty$ and $\|\cdot\|_{(n)} = \|\cdot\|_1$.

Let $\|\cdot\|$ be a given norm on M_n. The *dual norm* of $\|\cdot\|$ with respect to the Frobenius inner product is defined by

$$\|A\|^D \equiv \max\{|\operatorname{tr} AB^*| : \|B\| = 1; \ B \in M_n\}.$$

It is easy to verify that the dual norm of a unitarily invariant norm is also unitarily invariant. By Corollary 2.8, the dual norm of a unitarily invariant norm can be written as

$$\|A\|^D = \max\{\sum_{j=1}^{n} s_j(A)s_j(B) : \|B\| = 1, \ B \in M_n\}. \tag{4.1}$$

Thus we have $(\|\cdot\|^D)^D = \|\cdot\|$ by the duality theorem [51, Theorem 5.5.14].

Given $\gamma = (\gamma_1, \dots, \gamma_n)$ with $\gamma_1 \ge \cdots \ge \gamma_n \ge 0$, $\gamma_1 > 0$, we define

$$\|A\|_\gamma \equiv \sum_{j=1}^{n} \gamma_j s_j(A), \quad A \in M_n.$$

Then this is a unitarily invariant norm. Using this notation and considering the preceding remark, from (4.1) we get the following

Lemma 4.1 *Let $\|\cdot\|$ be a unitarily invariant norm on M_n and denote $\Gamma = \{s(X) : \|X\|^D = 1, \ X \in M_n\}$. Then for all A*

$$\|A\| = \max\{\|A\|_\gamma : \gamma \in \Gamma\}.$$

From this lemma and the formula

$$\|A\|_\gamma = \sum_{j=1}^{n-1} (\gamma_j - \gamma_{j+1})\|A\|_{(j)} + \gamma_n \|A\|_{(n)}$$

follows the next important

Lemma 4.2 (Fan Dominance Principle) *Let $A, B \in M_n$. If*

$$\|A\|_{(k)} \le \|B\|_{(k)} \quad \text{for } k = 1, 2, \dots, n$$

then

$$\|A\| \le \|B\|$$

for all unitarily invariant norms.

4.1 Operator Monotone Functions

In this section we derive some norm inequalities involving operator monotone functions and give applications.

We need the following fact [17, p.24].

Lemma 4.3 *Let $A \in M_n$ be Hermitian. Then for $1 \le k \le n$*

$$\sum_{j=1}^{k} \lambda_j(A) = \max \sum_{j=1}^{k} \langle Ax_j, x_j \rangle$$

where the maximum is taken over all orthonormal k-tuples $(x_1, \ldots, x_k)$ in $\mathbb{C}^n$.

By Theorem 1.3, every nonnegative operator monotone function $f(t)$ on $[0, \infty)$ has the following integral representation

$$f(t) = \alpha + \beta t + \int_0^\infty \frac{st}{s+t} d\mu(s) \tag{4.2}$$

where $\alpha, \beta \ge 0$ are constants and μ is a positive measure on $[0, \infty)$.

In this case, for any $A \ge 0$ and any vector u

$$\langle f(A)u, u \rangle = \alpha \langle u, u \rangle + \beta \langle Au, u \rangle + \int_0^\infty s \langle A(A+sI)^{-1}u, u \rangle d\mu(s). \tag{4.3}$$

Since $f(t)$ is increasing, for $j = 1, 2, \ldots, n$, a unit eigenvector u_j of A corresponding to $\lambda_j(A)$ becomes a unit eigenvector of $f(A)$ corresponding to $\lambda_j(f(A)) = f(\lambda_j(A))$, so that by the definition of Fan norms

$$\|f(A)\|_{(k)} = \sum_{j=1}^{k} \langle f(A)u_j, u_j \rangle, \quad k = 1, 2, \ldots, n. \tag{4.4}$$

Our first result is the following

Theorem 4.4 *Let A, B be positive semidefinite matrices and $\|\cdot\|$ be any unitarily invariant norm. Then the following assertions hold.*

(I) For every nonnegative operator monotone function $f(t)$ on $[0, \infty)$

$$\|f(A+B)\| \le \|f(A) + f(B)\|. \tag{4.5}$$

(II) For every nonnegative function $g(t)$ on $[0, \infty)$ with $g(0) = 0$ and $g(\infty) = \infty$, whose inverse function is operator monotone

$$\|g(A+B)\| \ge \|g(A) + g(B)\|. \tag{4.6}$$

The main part of the proof is to show successively two lemmas.

Now we denote by $\|X\|_F$ the Frobenius norm (i.e., the Schatten 2-norm) of a rectangular $l \times m$ matrix $X = (x_{ij})$:

$$\|X\|_F \equiv \{\mathrm{tr}\,(X^*X)\}^{1/2} = \{\mathrm{tr}\,(XX^*)\}^{1/2} = \{\sum_{i,j}|x_{ij}|^2\}^{1/2}.$$

Obviously the Frobenius norm is unitarily invariant in the sense that

$$\|UXV\|_F = \|X\|_F$$

for all X and all unitary matrices $U \in M_l$ and $V \in M_m$.

Lemma 4.5 *Given a matrix $C \geq 0$, let $\lambda_1 \geq \lambda_2 \geq \cdots \geq \lambda_n$ be its eigenvalues with corresponding orthonormal eigenvectors $v_1, v_2, \ldots, v_n$. Take any integer k with $1 \leq k \leq n$ and define an $n \times k$ matrix U_1 and an $n \times (n-k)$ matrix U_2 as*

$$U_1 \equiv (v_n, v_{n-1}, \ldots, v_{n-k+1}) \ \text{ and } \ U_2 \equiv (v_{n-k}, v_{n-k-1}, \ldots, v_1).$$

Then for every Hermitian matrix H

$$\|HCU_1\|_F \leq \|CHU_1\|_F \quad \text{and} \quad \|HCU_2\|_F \geq \|CHU_2\|_F.$$

Proof. Let

$$D_1 \equiv \mathrm{diag}(\lambda_n, \lambda_{n-1}, \ldots, \lambda_{n-k+1}),$$

$$D_2 \equiv \mathrm{diag}(\lambda_{n-k}, \lambda_{n-k-1}, \ldots, \lambda_1).$$

Then by definition we have

$$CU_1 = U_1 D_1 \quad \text{and} \quad CU_2 = U_2 D_2.$$

Since the matrix $W \equiv (U_1, U_2)$ and its adjoint W^* are unitary, we have

$$\begin{aligned}
\|HCU_1\|_F^2 &= \|W^*HU_1D_1\|_F^2 \\
&= \left\|\begin{bmatrix} U_1^*HU_1D_1 \\ U_2^*HU_1D_1 \end{bmatrix}\right\|_F^2 \\
&= \|U_1^*HU_1D_1\|_F^2 + \|U_2^*HU_1D_1\|_F^2 \\
&\leq \|U_1^*HU_1D_1\|_F^2 + \lambda_{n-k+1}^2\|U_2^*HU_1\|_F^2
\end{aligned}$$

where we have used the fact

$$\begin{aligned}
(U_2^*HU_1D_1)(U_2^*HU_1D_1)^* &= (U_2^*HU_1)D_1^2(U_2^*HU_1)^* \\
&\leq \lambda_{n-k+1}^2(U_2^*HU_1)(U_2^*HU_1)^*.
\end{aligned}$$

In a similar way we have

$$
\begin{aligned}
\|CHU_1\|_F^2 &= \|(CHU_1)^*W\|_F^2 \\
&= \|U_1^*HC \cdot (U_1, U_2)\|_F^2 \\
&= \|[U_1^*HU_1D_1, U_1^*HU_2D_2]\|_F^2 \\
&= \|U_1^*HU_1D_1\|_F^2 + \|U_1^*HU_2D_2\|_F^2 \\
&= \|U_1^*HU_1D_1\|_F^2 + \|D_2U_2^*HU_1\|_F^2 \\
&\geq \|U_1^*HU_1D_1\|_F^2 + \lambda_{n-k}^2\|U_2^*HU_1\|_F^2.
\end{aligned}
$$

Combining these two inequalities yields the first inequality of the assertion. A similar consideration proves the second inequality. This completes the proof.

$\square$

Lemma 4.6 *Let $A, B \geq 0$ and let u_j be the orthonormal eigenvectors of $A+B$ corresponding to $\lambda_j(A + B)$, $j = 1, 2, \ldots, n$. Then the following inequalities hold:*

$$
\sum_{j=1}^{k}\langle\{A(A + I)^{-1} + B(B + I)^{-1}\}u_j, u_j\rangle
$$

$$
\geq \sum_{j=1}^{k}\langle(A + B)(A + B + I)^{-1}u_j, u_j\rangle
$$

for $k = 1, 2, \ldots, n$.

Proof. Let $C \equiv (A + B + I)^{-1/2}$ and denote its eigenvalues by $\lambda_1 \geq \lambda_2 \geq \cdots \geq \lambda_n$. Since

$$
(A + B)(A + B + I)^{-1} = CAC + CBC,
$$

for the proof of the lemma it suffices to prove the following two inequalities:

$$
\sum_{j=1}^{k}\langle\{A(A + I)^{-1}\}u_j, u_j\rangle \geq \sum_{j=1}^{k}\langle CACu_j, u_j\rangle \tag{4.7}
$$

and

$$
\sum_{j=1}^{k}\langle\{B(B + I)^{-1}\}u_j, u_j\rangle \geq \sum_{j=1}^{k}\langle CBCu_j, u_j\rangle. \tag{4.8}
$$

For each $j = 1, 2, \ldots, n$ the vector u_j coincides with the eigenvector v_{n-j+1} of C corresponding to λ_{n-j+1}. Let

$$
U_1 \equiv (v_n, v_{n-1}, \ldots, v_{n-k+1}) = (u_1, u_2, \ldots, u_k).
$$

Then the inequality (4.7) can be written as

$$\operatorname{tr}\left[U_1^* A(A+I)^{-1}U_1\right] \geq \operatorname{tr}\left(U_1^* CACU_1\right). \tag{4.9}$$

Applying Lemma 4.5 with $A^{1/2}$ in place of H, and using the obvious inequality

$$C^2 = (A+B+I)^{-1} \leq (A+I)^{-1}$$

we obtain

$$
\begin{aligned}
\operatorname{tr}\left(U_1^* CACU_1\right) &= \|A^{1/2}CU_1\|_F^2 \\
&\leq \|CA^{1/2}U_1\|_F^2 = \operatorname{tr}\left(U_1^* A^{1/2}C^2 A^{1/2}U_1\right) \\
&\leq \operatorname{tr}\left(U_1^* A^{1/2}(A+I)^{-1}A^{1/2}U_1\right) \\
&= \operatorname{tr}\left(U_1^* A(A+I)^{-1}U_1\right),
\end{aligned}
$$

which proves (4.9). Finally (4.8) follows from (4.7) via interchange of the roles of A and B. This completes the proof. $\qquad\square$

Proof of Theorem 4.4. It is easy to see from Lemma 4.6 that for every $s > 0$

$$\sum_{j=1}^{k} s\langle (A+B)(A+B+sI)^{-1}u_j,\, u_j\rangle$$

$$\leq \sum_{j=1}^{k} \langle \{sA(A+sI)^{-1} + sB(B+sI)^{-1}\}u_j,\, u_j\rangle.$$

Then by the representation (4.3) we have

$$\sum_{j=1}^{k} \langle f(A+B)u_j,\, u_j\rangle \leq \sum_{j=1}^{k} \langle \{f(A) + f(B)\}u_j,\, u_j\rangle,$$

and by (4.4) applied to $A+B$ in place of A,

$$\sum_{j=1}^{k} \langle f(A+B)u_j,\, u_j\rangle = \|f(A+B)\|_{(k)}.$$

On the other hand, Lemma 4.3 shows

$$\sum_{j=1}^{k} \langle \{f(A)+f(B)\}u_j,\, u_j\rangle \leq \|f(A)+f(B)\|_{(k)}$$

and consequently

$$\|f(A+B)\|_{(k)} \leq \|f(A)+f(B)\|_{(k)}$$

for $k = 1, 2, \ldots, n$. Finally apply the Fan dominance principle (Lemma 4.2) to complete the proof of (I).

To prove (II), apply (I) to the inverse function $f(t)$ of $g(t)$, which is operator monotone, matrices $g(A)$, $g(B) \geq 0$, and the Fan norms to get

$$\|A + B\|_{(k)} = \|f[g(A)] + f[g(B)]\|_{(k)}$$
$$\geq \|f[g(A) + g(B)]\|_{(k)}.$$

Thus

$$\|A + B\|_{(k)} \geq \|f[g(A) + g(B)]\|_{(k)} \quad \text{for } k = 1, 2, \ldots, n. \tag{4.10}$$

Since a nonnegative continuous function defined on $[0, \infty)$ is operator monotone if and only if it is operator concave [17, Theorem V.2.5], $f(t)$ is an increasing concave function. Hence $g(t)$ is an increasing convex function. Applying Theorem 2.3 with $g(t)$ to (4.10) we get

$$\|g(A + B)\|_{(k)} \geq \|g(A) + g(B)\|_{(k)} \quad \text{for} \quad k = 1, 2, \ldots, n$$

which implies, by Lemma 4.2,

$$\|g(A + B)\| \geq \|g(A) + g(B)\|$$

for all unitarily invariant norms. This completes the proof. $\qquad\square$

Corollary 4.7 *For all $X, Y \geq 0$ and every unitarily invariant norm $\| \cdot \|$,*

$$\|(X + Y)^r\| \leq \|X^r + Y^r\| \quad (0 < r \leq 1) \tag{4.11}$$

and

$$\|(X + Y)^r\| \geq \|X^r + Y^r\| \quad (1 \leq r < \infty). \tag{4.12}$$

Proof. Since t^r is operator monotone for $0 < r \leq 1$ and the inverse function of t^r is operator monotone for $1 \leq r < \infty$, applying Theorem 4.4 completes the proof. $\qquad\square$

Corollary 4.7 will be used in the proofs of Theorems 4.16 and 4.35 below.

Corollary 4.8 *For all $A, B \geq 0$ and every unitarily invariant norm $\| \cdot \|$,*

$$\| \log(A + B + I)\| \leq \| \log(A + I) + \log(B + I)\|$$

and

$$\|e^A + e^B\| \leq \|e^{A+B} + I\|.$$

Proof. The first inequality follows from Theorem 4.4, applied to the operator monotone function $\log(t+1)$. Again by Theorem 4.4, applied to the Fan norms and the function $e^t - 1$ whose inverse function is operator monotone, we have for every $1 \le k \le n$

$$\|(e^A - I) + (e^B - I)\|_{(k)} \le \|e^{A+B} - I\|_{(k)}.$$

Since by the definition of Fan norms

$$\|(e^A - I) + (e^B - I)\|_{(k)} = \|e^A + e^B\|_{(k)} - 2k$$

and

$$\|e^{A+B} - I\|_{(k)} = \|e^{A+B} + I\|_{(k)} - 2k,$$

we can conclude

$$\|e^A + e^B\|_{(k)} \le \|e^{A+B} + I\|_{(k)}$$

for every $1 \le k \le n$. Now the Fan dominance principle yields the second inequality. $\square$

Corollary 4.9 *Let $g(t)$ be a nonnegative operator convex function on $[0, \infty)$ with $g(0) = 0$. Then*

$$\|g(A + B)\| \ge \|g(A) + g(B)\|$$

for all A, $B \ge 0$ and every unitarily invariant norm.

Proof. It is known [17, Theorem V.2.9] that a function $g(t)$ on $[0, \infty)$ with $g(0) = 0$ is operator convex if and only if $g(t)/t$ is operator monotone. Therefore we may assume that $g(t) = tf(t)$ with $f(t)$ being operator monotone. It is proved in [4, Lemma 5] that the inverse function of $tf(t)$ is operator monotone. Thus applying Theorem 4.4(II) completes the proof. $\square$

We remark that the following result on matrix differences was proved earlier by T. Ando [4] : Under the same conditions as in Theorem 4.4,

$$\|f(|A - B|)\| \ge \|f(A) - f(B)\|,$$

$$\|g(|A - B|)\| \le \|g(A) - g(B)\|$$

which can be found in [17, Section X.1].

A norm on M_n is said to be *normalized* if $\|\mathrm{diag}(1, 0, \ldots, 0)\| = 1$. Evidently all the Fan k-norms ($k = 1, \ldots, n$) and Schatten p-norms ($1 \le p \le \infty$) are normalized.

If $\| \cdot \|$ is a unitarily invariant norm on M_n and $A \ge 0$, then Lemma 4.1 has the following equivalent form:

$$\|A\| = \max\{\mathrm{tr}\, AB : B \ge 0, \|B\|^D = 1, B \in M_n\}. \tag{4.13}$$

Theorem 4.10 *Let $f(t)$ be a nonnegative operator monotone function on $[0, \infty)$ and $\| \cdot \|$ be a normalized unitarily invariant norm. Then for every matrix A,*

$$f(\|A\|) \le \|f(|A|)\|. \tag{4.14}$$

Proof. Since $\|A\| = \|\,|A|\,\|$, it suffices to prove (4.14) for the case when A is positive semidefinite. Now we make this assumption. By (4.13) there exists a matrix $B \ge 0$ with $\|B\|^D = 1$ such that

$$\|A\| = \operatorname{tr} AB. \tag{4.15}$$

It is easy to see [17, p.93] that every normalized unitarily invariant norm satisfies

$$\|X\|_\infty \le \|X\| \le \|X\|_1 \quad \text{for all } X.$$

Since $\|\cdot\|$ is normalized, $\|\cdot\|^D$ is also a normalized unitarily invariant norm. Hence

$$1 = \|B\|^D \le \|B\|_1 = \operatorname{tr} B. \tag{4.16}$$

From $\|A\|_\infty \le \|A\|$ and (4.15) we have

$$\frac{s\|A\|}{s+\|A\|} \le \frac{s\|A\|}{s+\|A\|_\infty} = \operatorname{tr} \frac{sAB}{s+\|A\|_\infty}$$

$$= \operatorname{tr} B^{1/2} \frac{sA}{s+\|A\|_\infty} B^{1/2}$$

$$\le \operatorname{tr} B^{1/2} \{sA(sI+A)^{-1}\} B^{1/2}$$

$$= \operatorname{tr} sA(sI+A)^{-1} B$$

for any real number $s > 0$. In the above latter inequality we have used the fact that $sA/(s+\|A\|_\infty) \le sA(sI+A)^{-1}$. Thus

$$\frac{s\|A\|}{s+\|A\|} \le \operatorname{tr} sA(sI+A)^{-1}B. \tag{4.17}$$

Using the integral representation (4.2) of $f(t)$, (4.16), (4.15), (4.17) and (4.13) we compute

$$f(\|A\|) = \alpha + \beta\|A\| + \int_0^\infty \frac{s\|A\|}{s+\|A\|}\, d\mu(s)$$

$$\le \alpha\operatorname{tr} B + \beta\operatorname{tr} AB + \int_0^\infty \operatorname{tr} sA(sI+A)^{-1}B\, d\mu(s)$$

$$= \operatorname{tr} \left\{ \alpha I + \beta A + \int_0^\infty sA(sI+A)^{-1} d\mu(s) \right\} B$$

$$= \operatorname{tr} f(A)B$$

$$\le \|f(A)\|.$$

This completes the proof.   □

We say that a norm $\|\cdot\|$ is *strictly increasing* if $0 \leq A \leq B$ and $\|A\| = \|B\|$ imply $A = B$. For instance, the Schatten p-norm $\|\cdot\|_p$ is strictly increasing for all $1 \leq p < \infty$. Now we consider the equality case of (4.14).

Theorem 4.11 *Let $f(t)$ be a nonnegative operator monotone function on $[0, \infty)$ and assume that $f(t)$ is non-linear. Let $\|\cdot\|$ be a strictly increasing normalized unitarily invariant norm and $A \in M_n$ with $n \geq 2$. Then $f(\|A\|) = \|f(|A|)\|$ if and only if $f(0) = 0$ and rank $A \leq 1$.*

Proof. First assume that $f(0) = 0$ and $|A| = \lambda P$ with P a projection of rank one. Then $\|A\| = \lambda\|P\| = \lambda$ by the normalization assumption and $\|f(|A|)\| = \|f(\lambda)P\| = f(\lambda) = f(\|A\|)$.

Conversely, assume $f(\|A\|) = \|f(|A|)\|$. If $A = 0$, then since $\|\cdot\|$ is normalized and strictly increasing we must have $f(0) = 0$. Next suppose $A \neq 0$. Let μ be the measure in the integral representation (4.2) of $f(t)$. Since $f(t)$ is non-linear, $\mu \neq 0$. From the proof of Theorem 4.10 we know that $f(\|A\|) = \|f(|A|)\|$ implies $\|A\|_\infty = \|A\|$ or equivalently

$$\|\mathrm{diag}(s_1, 0, \ldots, 0)\| = \|\mathrm{diag}(s_1, s_2, \ldots, s_n)\|$$

where $s_1 \geq s_2 \geq \cdots \geq s_n$ are the singular values of A. Now the strict increasingness of $\|\cdot\|$ forces $s_2 = \cdots = s_n = 0$, that is, rank $A = 1$. So write $|A| = \lambda P$ with P a projection of rank one. Since $f(\|A\|) = \|f(|A|)\|$ means

$$\|f(\lambda)P\| = \|f(\lambda)P + f(0)(I - P)\|,$$

we have $f(0) = 0$ due to $I - P \neq 0$ and the strict increasingness of $\|\cdot\|$ again. This completes the proof.   □

Theorem 4.10 can be complemented by the following reverse inequality for unitarily invariant norms with a different normalization.

Theorem 4.12 *Let $f(t)$ be a nonnegative operator monotone function on $[0, \infty)$ and $\|\cdot\|$ be a unitarily invariant norm with $\|I\| = 1$. Then for every matrix A,*

$$f(\|A\|) \geq \|f(|A|)\|.$$

Proof. We may assume that A is positive semidefinite. Since

$$f(A) = \alpha I + \beta A + \int_0^\infty sA(sI + A)^{-1}\, d\mu(s)$$

as in the proof of Theorem 4.10, we have

$$\|f(A)\| \leq \alpha + \beta\|A\| + \int_0^\infty \|sA(sI + A)^{-1}\|\, d\mu(s)$$

due to $\|I\| = 1$. Hence it suffices to show

$$\|A(sI + A)^{-1}\| \le \frac{\|A\|}{s + \|A\|} \qquad (s > 0). \qquad (4.18)$$

For each $s > 0$, since

$$\frac{x}{s + x} \le t^2 + (1 - t)^2 \frac{x}{s}$$

for all $x > 0$ and $0 < t < 1$, we get

$$A(sI + A)^{-1} \le t^2 I + (1 - t)^2 s^{-1} A$$

so that

$$\|A(sI + A)^{-1}\| \le \|t^2 I + (1 - t)^2 s^{-1} A\| \le t^2 + (1 - t)^2 s^{-1} \|A\|. \qquad (4.19)$$

Minimize the right-hand side of (4.19) over $t \in (0, 1)$ to obtain (4.18). This completes the proof. □

Denote $E \equiv \mathrm{diag}(1, 0, \dots, 0)$. Combining Theorems 4.10 and 4.12, we get the following two corollaries.

Corollary 4.13 *Let $f(t)$ be a nonnegative operator monotone function on $[0, \infty)$ and $\| \cdot \|$ be a unitarily invariant norm. Then for every matrix A,*

$$\|E\| \cdot f\left(\frac{\|A\|}{\|E\|}\right) \le \|f(|A|)\| \le \|I\| \cdot f\left(\frac{\|A\|}{\|I\|}\right).$$

As an immediate consequence of the above corollary we have the next

Corollary 4.14 *Let $g(t)$ be a strictly increasing function on $[0, \infty)$ such that $g(0) = 0$, $g(\infty) = \infty$ and its inverse function g^{-1} on $[0, \infty)$ is operator monotone. Let $\| \cdot \|$ be a unitarily invariant norm. Then for every matrix A,*

$$\|I\| \cdot g\left(\frac{\|A\|}{\|I\|}\right) \le \|g(|A|)\| \le \|E\| \cdot g\left(\frac{\|A\|}{\|E\|}\right).$$

Next we study monotonicity properties of some norm functions by applying several special cases of Theorems 4.4 and 4.10.

Given a unitarily invariant norm $\| \cdot \|$ on M_n, for $p > 0$ define

$$\|X\|^{(p)} \equiv \| |X|^p \|^{1/p} \qquad (X \in M_n). \qquad (4.20)$$

Then it is known [17, p.95] (or [46, Lemma 2.13]) that when $p \ge 1$, $\| \cdot \|^{(p)}$ is also a unitarily invariant norm.

Theorem 4.15 *Let $\| \cdot \|$ be a normalized unitarily invariant norm. Then for any matrix A, the function $p \mapsto \|A\|^{(p)}$ is decreasing on $(0, \infty)$. For any unitarily invariant norm*

$$\lim_{p \to \infty} \|A\|^{(p)} = \|A\|_\infty.$$

Proof. The monotonicity part is the special case of Theorem 4.10 when $f(t) = t^r$, $0 < r \le 1$, but we may give a short direct proof. It suffices to consider the case when A is positive semidefinite, and now we make this assumption. First assume that $\| \cdot \|$ is normalized. Let us show

$$\|A^r\| \ge \|A\|^r \qquad (0 < r \le 1), \qquad (4.21)$$

$$\|A^r\| \le \|A\|^r \qquad (1 \le r < \infty). \qquad (4.22)$$

Since $\|A\|_\infty \le \|A\|$, for $r \ge 1$ we get

$$\|A^r\| = \|AA^{r-1}\| \le \|A\| \, \|A^{r-1}\|_\infty$$
$$= \|A\| \, \|A\|_\infty^{r-1} \le \|A\| \, \|A\|^{r-1} = \|A\|^r,$$

proving (4.22). The inequality (4.21) follows from (4.22): For $0 < r \le 1$, $\|A\| = \|(A^r)^{1/r}\| \le \|A^r\|^{1/r}$.

If $0 < p < q$, then

$$\|A^p\| = \|(A^q)^{p/q}\| \ge \|A^q\|^{p/q}$$

so that $\|A^p\|^{1/p} \ge \|A^q\|^{1/q}$. Moreover,

$$\|A\|_\infty = \|A^p\|_\infty^{1/p} \le \|A^p\|^{1/p} \le \|A^p\|_1^{1/p} = \|A\|_p \to \|A\|_\infty$$

as $p \to \infty$, where $\|\cdot\|_p$ is the Schatten p-norm. For the limit assertion when $\|\cdot\|$ is not normalized, we just apply the normalized case to $\|\cdot\|/\|\mathrm{diag}(1, 0, \dots, 0)\|$. $\qquad \square$

Next we consider the monotonicity of the functions $p \mapsto \|(A^p + B^p)^{1/p}\|$ and $p \mapsto \|A^p + B^p\|^{1/p}$. We denote by $A \vee B$ the supremum of two positive semidefinite matrices A, B in the sense of Kato [60] (see also [5, Lemma 6.15]); it is the limit of an increasing sequence: $A \vee B = \lim_{p \to \infty} \{(A^p + B^p)/2\}^{1/p}$.

Theorem 4.16 Let $A, B \in M_n$ be positive semidefinite. For every unitarily invariant norm, the function $p \mapsto \|(A^p + B^p)^{1/p}\|$ is decreasing on $(0, 1]$. For every normalized unitarily invariant norm, the function $p \mapsto \|A^p + B^p\|^{1/p}$ is decreasing on $(0, \infty)$ and for every unitarily invariant norm

$$\lim_{p \to \infty} \|A^p + B^p\|^{1/p} = \|A \vee B\|_\infty.$$

Proof. Let $0 < p < q \le 1$. Set $r = q/p$ (> 1), $X = A^p$, $Y = B^p$ in (4.12) to get

$$\|(A^p + B^p)^{q/p}\| \ge \|A^q + B^q\|. \qquad (4.23)$$

Using a majorization principle (Theorem 2.3) and Fan's dominance principle (Lemma 4.2), we can apply the increasing convex function $t^{1/q}$ on $[0, \infty)$ to (4.23) and get

$$\|(A^p + B^p)^{1/p}\| \geq \|(A^q + B^q)^{1/q}\|$$

which shows the first assertion.

To show the second assertion we must prove

$$\|A^p + B^p\|^{1/p} \geq \|A^q + B^q\|^{1/q} \qquad (4.24)$$

for $0 < p < q$ and every normalized unitarily invariant norm $\|\cdot\|$. It is easily seen that (4.24) is equivalent to

$$\|A + B\|^r \geq \|A^r + B^r\|$$

for all $r \geq 1$ and all positive semidefinite $A, B \in M_n$, which follows from (4.22) and (4.12):

$$\|A + B\|^r \geq \|(A + B)^r\| \geq \|A^r + B^r\|.$$

Again, to show the limit assertion it suffices to consider the case when $\|\cdot\|$ is normalized. In this case for $p \geq 1$,

$$
\begin{aligned}
\|(A^p + B^p)^{1/p}\|_\infty &= \|A^p + B^p\|_\infty^{1/p} \\
&\leq \|A^p + B^p\|^{1/p} \\
&\leq \|A^p + B^p\|_1^{1/p} \\
&= \|(A^p + B^p)^{1/p}\|_p \\
&\leq \|(A^p + B^p)^{1/p} - (A \vee B)\|_p + \|A \vee B\|_p \\
&\leq \|(A^p + B^p)^{1/p} - (A \vee B)\|_1 + \|A \vee B\|_p.
\end{aligned}
$$

Since

$$\lim_{p \to \infty} (A^p + B^p)^{1/p} = \lim_{p \to \infty} \left(\frac{A^p + B^p}{2}\right)^{1/p} = A \vee B$$

and

$$\lim_{p \to \infty} \|A \vee B\|_p = \|A \vee B\|_\infty,$$

we obtain

$$\lim_{p \to \infty} \|A^p + B^p\|^{1/p} = \|A \vee B\|_\infty.$$

This completes the proof. $\square$

We remark that there are some unitarily invariant norms for which $p \mapsto \|(A^p + B^p)^{1/p}\|$ is not decreasing on $(1, \infty)$. Consider the trace norm (here it is just trace since the matrices involved are positive semidefinite). In fact, for the 2×2 matrices

$$A = \begin{bmatrix} 1 & 0 \\ 0 & 0 \end{bmatrix}, \quad B_t = \begin{bmatrix} t^2 & t\sqrt{1-t^2} \\ t\sqrt{1-t^2} & 1-t^2 \end{bmatrix} \quad (0 < t < 1),$$

it is proved in [10, Lemma 3.3] that for any $p_0 > 2$ there exists a $t \in (0,1)$ such that $p \mapsto \mathrm{tr}\left\{(A^p + B_t^p)^{1/p}\right\}$ is strictly increasing on $[p_0, \infty)$. Also consider the example $\psi(p) = \mathrm{tr}\left\{(A^p + B^p)^{1/p}\right\}$ with

$$A = \begin{bmatrix} 1 & 2 \\ 2 & 5 \end{bmatrix}, \quad B = \begin{bmatrix} 1 & -6 \\ -6 & 50 \end{bmatrix}.$$

Then $\psi(1.5) - \psi(8) \approx -1.5719$. Thus $\psi(1.5) < \psi(8)$. Hence $\psi(p)$ is not decreasing on the interval $[1.5, 8]$.

Notes and References Theorem 4.4 and its three corollaries are proved by T. Ando and X. Zhan [11]. Part (I) of Theorem 4.4 is conjectured by F. Hiai [46] as well as by R. Bhatia and F. Kittaneh [23]. The other results in this section are proved by F. Hiai and X. Zhan [49].

4.2 Cartesian Decompositions Revisited

In Section 3.4 by establishing majorization relations for singular values we have derived some Schatten p-norm inequalities associated with the matrix Cartesian decomposition. Now let us prove an inequality for generic unitarily invariant norms.

For $x = (x_j) \in \mathbb{R}^n$ define

$$\Phi_k(x) \equiv \max\left\{\sum_{m=1}^{k} |x_{j_m}| : 1 \le j_1 < j_2 < \cdots < j_k \le n\right\}.$$

Then $\Phi_{(k)}(\cdot)$ is the symmetric gauge function corresponding to the Fan k-norm $\|\cdot\|_{(k)}$:

$$\|A\|_{(k)} = \Phi_{(k)}(s(A)).$$

We need the following useful fact.

Lemma 4.17 *Let $T \in M_n$. Then for each $k = 1, 2, \ldots, n$*

$$\|T\|_{(k)} = \min\{\|X\|_1 + k\|Y\|_\infty : T = X + Y\}.$$

Proof. If $T = X + Y$, then

$$\|T\|_{(k)} \le \|X\|_{(k)} + \|Y\|_{(k)} \le \|X\|_1 + k\|Y\|_\infty.$$

Now let $A = U\mathrm{diag}(s_1, \ldots, s_n)V$ be the singular value decomposition with U, V unitary and $s_1 \ge \cdots \ge s_n \ge 0$. Then

$$X \equiv U \mathrm{diag}(s_1 - s_k, s_2 - s_k, \ldots, s_k - s_k, 0, \ldots, 0)V$$

and

$$Y \equiv U \mathrm{diag}(s_k, \ldots, s_k, s_{k+1}, \ldots, s_n)V$$

satisfy

$$T = X + Y$$

and

$$\|T\|_{(k)} = \|X\|_1 + k\|Y\|_\infty.$$

$\square$

For $x = (x_j) \in \mathbb{C}^n$ we denote $|x| = (|x_1|, \ldots, |x_n|)$. Note that the singular values of a Hermitian matrix are just the absolute values of its eigenvalues. In this section $i = \sqrt{-1}$.

Theorem 4.18 *Let $T = A + iB \in M_n$ with A, B Hermitian. Let α_j and β_j be the eigenvalues of A and B respectively ordered so that $|\alpha_1| \geq \cdots \geq |\alpha_n|$ and $|\beta_1| \geq \cdots \geq |\beta_n|$. Then*

$$\|\mathrm{diag}(\alpha_1 + i\beta_1, \ldots, \alpha_n + i\beta_n)\| \leq \sqrt{2}\,\|T\| \qquad (4.25)$$

for every unitarily invariant norm.

Proof. Observe that

$$\|\mathrm{diag}(\alpha_1 + i\beta_1, \ldots, \alpha_n + i\beta_n)\| = \|\mathrm{diag}(|\alpha_1| + i|\beta_1|, \ldots, |\alpha_n| + i|\beta_n|)\|$$

for every unitarily invariant norm.

By the Fan dominance principle (Lemma 4.2), it suffices to prove (4.25) for all Fan k-norms, $k = 1, 2, \ldots, n$. We first show that (4.25) is true for the two special cases when $\|\cdot\|$ is the trace norm $\|\cdot\|_{(n)}\ (=\|\cdot\|_1)$ or the spectral norm $\|\cdot\|_{(1)}\ (=\|\cdot\|_\infty)$. The case $p = 1$ of the inequality (3.37) in Section 3.4 shows that (4.25) is true for the trace norm. From

$$A = \frac{T + T^*}{2}, \qquad B = \frac{T - T^*}{2i}$$

we have $|\alpha_1| = \|A\|_{(1)} \leq \|T\|_{(1)}$, $|\beta_1| = \|B\|_{(1)} \leq \|T\|_{(1)}$, and hence $|\alpha_1 + i\beta_1| \leq \sqrt{2}\|T\|_{(1)}$. So (4.25) holds for the spectral norm.

Now let us fix k with $1 \leq k \leq n$. By Lemma 4.17 there exist $X, Y \in M_n$ such that $T = X + Y$ and

$$\|T\|_{(k)} = \|X\|_{(n)} + k\|Y\|_{(1)}. \qquad (4.26)$$

Let $X = C + iD$ and $Y = E + iF$ be the Cartesian decompositions of X and Y, i.e., C, D, E, F are all Hermitian. Thus $T = C + E + i(D + F)$. Since the Cartesian decomposition of a matrix is unique, we must have

$$A = C + E \quad \text{and} \quad B = D + F. \tag{4.27}$$

By the two already proved cases of (4.25) we get

$$\sqrt{2}\|X\|_{(n)} \geq \Phi_{(n)}(|s(C) + is(D)|) \tag{4.28}$$

and

$$\sqrt{2}\|Y\|_{(1)} \geq \Phi_{(1)}(|s(E) + is(F)|). \tag{4.29}$$

Combining (4.26), (4.28) and (4.29) we obtain

$$\sqrt{2}\|T\|_{(k)} \geq \Phi_{(n)}(|s(C) + is(D)|) + k\Phi_{(1)}(|s(E) + is(F)|)$$

$$\geq \sum_{j=1}^{k} |s_j(C) + is_j(D)| + k|s_1(E) + is_1(F)|.$$

Thus

$$\sqrt{2}\|T\|_{(k)} \geq \sum_{j=1}^{k} |s_j(C) + is_j(D)| + k|s_1(E) + is_1(F)|. \tag{4.30}$$

A well-known fact [17, p.99] says that for any $W, Z \in M_n$,

$$\max_j |s_j(W) - s_j(Z)| \leq s_1(W - Z).$$

So

$$s_j(C + E) \leq s_j(C) + s_1(E), \quad s_j(D + F) \leq s_j(D) + s_1(F) \tag{4.31}$$

for $j = 1, 2, \ldots, n$.

Using (4.31) and (4.27) we compute

$$\sum_{j=1}^{k} |s_j(C) + is_j(D)| + k|s_1(E) + is_1(F)|$$

$$= \sum_{j=1}^{k} \{|s_j(C) + is_j(D)| + |s_1(E) + is_1(F)|\}$$

$$\geq \sum_{j=1}^{k} |[s_j(C) + s_1(E)] + i[s_j(D) + s_1(F)]|$$

$$\geq \sum_{j=1}^{k} |s_j(C + E) + is_j(D + F)|$$

$$= \Phi_{(k)}(|s(A) + is(B)|).$$

Therefore

$$\sum_{j=1}^{k} |s_j(C) + is_j(D)| + k|s_1(E) + is_1(F)| \geq \Phi_{(k)}(|s(A) + is(B)|). \quad (4.32)$$

Combining (4.30) and (4.32) we obtain

$$\sqrt{2}\|T\|_{(k)} \geq \Phi_{(k)}(|s(A) + is(B)|), \quad k = 1, 2, \ldots, n.$$

This proves (4.25). □

We remark that the factor $\sqrt{2}$ in the inequality (4.25) is best possible. To see this consider the trace norm $\|\cdot\|_{(2)}$ with the example

$$A = \begin{bmatrix} 0 & 1 \\ 1 & 0 \end{bmatrix} \quad \text{and} \quad B = \begin{bmatrix} 0 & i \\ -i & 0 \end{bmatrix}.$$

A matrix S is called skew-Hermitian if $S^* = -S$. Let us arrange the eigenvalues λ_j of $G \in M_n$ in decreasing order of magnitude: $|\lambda_1| \geq \cdots \geq |\lambda_n|$ and denote $\text{Eig}(G) \equiv \text{diag}(\lambda_1, \ldots, \lambda_n)$. We can also interpret the inequality (4.25) as the following spectral variation (perturbation) theorem.

Theorem 4.18' *If H is Hermitian and S is skew-Hermitian, then*

$$\|\text{Eig}(H) - \text{Eig}(S)\| \leq \sqrt{2}\|H - S\|$$

for every unitarily invariant norm.

Notes and References. Theorem 4.18 is conjectured by T. Ando and R. Bhatia [9, p.142]; it is proved by X. Zhan [88]. The equivalent form Theorem 4.18' is a conjecture in R. Bhatia's book [16, p.119]. Lemma 4.17 can be found in [17, Proposition IV.2.3].

4.3 Arithmetic-Geometric Mean Inequalities

The arithmetic-geometric mean inequality for complex numbers a, b is $|ab| \leq (|a|^2 + |b|^2)/2$. One matrix version of this inequality is the following result.

Theorem 4.19 *For any three matrices A, B, X we have*

$$\|AXB^*\| \leq \frac{1}{2}\|A^*AX + XB^*B\| \quad (4.33)$$

for every unitarily invariant norm.

To give a proof we first make some preparations. For a matrix A we write its real part as $\text{Re}A = (A + A^*)/2$, while for a vector $x = (x_1, \ldots, x_n)$ we denote $\text{Re}x = (\text{Re}x_1, \ldots, \text{Re}x_n)$ and $|x| = (|x_1|, \ldots, |x_n|)$.

Lemma 4.20 (Ky Fan) *For every matrix A*

$$\text{Re}\lambda(A) \prec \lambda(\text{Re}A).$$

Proof. Since any matrix is unitarily equivalent to an upper triangular matrix [51, p.79], for our problem we may assume that A is upper triangular. Then the components of $\text{Re}\lambda(A)$ are the diagonal entries of $\text{Re}A$. Hence applying Theorem 2.1 yields the conclusion.   □

Lemma 4.21 *Let A, B be any two matrices such that the product AB is Hermitian. Then*

$$\|AB\| \le \|\text{Re}(BA)\|$$

for every unitarily invariant norm.

Proof. Since $\lambda(BA) = \lambda(AB)$, the eigenvalues of BA are all real. By Lemma 4.20,

$$\lambda(BA) \prec \lambda(\text{Re}(BA)). \tag{4.34}$$

Applying Theorem 2.2 to the convex function $f(t) = |t|$ on $\mathbb{R}$, we know that $x \prec y$ implies $|x| \prec_w |y|$. Thus from (4.34) we get

$$|\lambda(AB)| = |\lambda(BA)| \prec_w |\lambda[\text{Re}(BA)]|.$$

Since AB and $\text{Re}(BA)$ are Hermitian, this is the same as

$$s(AB) \prec_w s[\text{Re}(BA)].$$

Now Fan's dominance principle gives

$$\|AB\| \le \|\text{Re}(BA)\|,$$

completing the proof.   □

First Proof of Theorem 4.19. By the polar decompositions $A = U|A|$, $B = V|B|$ with U, V unitary, it suffices to prove (4.33) for the case when A, B are positive semidefinite. Now we make this assumption. First consider the case when $A = B$ and $X^* = X$. Then Lemma 4.21 yields

$$\|AXA\| \le \|\text{Re}(XA^2)\| = \frac{1}{2}\|A^2X + XA^2\|. \tag{4.35}$$

Next consider the general case. Let

$$T = \begin{bmatrix} A & 0 \\ 0 & B \end{bmatrix}, \quad Y = \begin{bmatrix} 0 & X \\ X^* & 0 \end{bmatrix}.$$

Replacing A and X in (4.35) by T and Y respectively, we get

$$\left\| \begin{bmatrix} 0 & AXB \\ (AXB)^* & 0 \end{bmatrix} \right\| \leq \frac{1}{2} \left\| \begin{bmatrix} 0 & A^2X + XB^2 \\ (A^2X + XB^2)^* & 0 \end{bmatrix} \right\|.$$

Note that for $C \in M_n$,

$$s\left(\begin{bmatrix} 0 & C \\ C^* & 0 \end{bmatrix} \right) = (s_1(C), s_1(C), s_2(C), s_2(C), \dots, s_n(C), s_n(C)).$$

Using Fan's dominance principle it is clear that the above inequality holds for all unitarily invariant norms if and only if

$$\|AXB\| \leq \frac{1}{2}\|A^2X + XB^2\|$$

holds for all unitarily invariant norms. This completes the proof. $\square$

For $0 < \theta < 1$, we set $d\mu_\theta(t) = a_\theta(t)dt$ and $d\nu_\theta(t) = b_\theta(t)dt$ where

$$a_\theta(t) = \frac{\sin(\pi\theta)}{2(\cosh(\pi t) - \cos(\pi\theta))}, \quad b_\theta(t) = \frac{\sin(\pi\theta)}{2(\cosh(\pi t) + \cos(\pi\theta))}.$$

For a bounded continuous function $f(z)$ on the strip $\Omega = \{z \in \mathbb{C} : 0 \leq \mathrm{Im}z \leq 1\}$ which is analytic in the interior, we have the well-known Poisson integral formula (see e.g. [81])

$$f(i\theta) = \int_{-\infty}^{\infty} f(t)d\mu_\theta(t) + \int_{-\infty}^{\infty} f(i+t)d\nu_\theta(t), \qquad (4.36)$$

and the total masses of the measures $d\mu_\theta(t)$, $d\nu_\theta(t)$ are $1 - \theta, \theta$ respectively [64, Appendix B]. In particular, $d\mu_{1/2}(t) = d\nu_{1/2}(t) = dt/2\cosh(\pi t)$ has the total mass $1/2$. Now we use this integral formula to give another proof of the inequality (4.33).

Second Proof of Theorem 4.19. The inequality (4.33) is equivalent to

$$\|A^{1/2}XB^{1/2}\| \leq \frac{1}{2}\|AX + XB\| \qquad (4.37)$$

for $A, B \geq 0$. Further by continuity we may assume that A, B are positive definite. Then the function

$$f(t) = A^{1+it}XB^{-it} \qquad (t \in \mathbb{R})$$

extends to a bounded continuous function on the strip Ω which is analytic in the interior. Here the matrices A^{it} and B^{-it} are unitary. By (4.36) we have

$$A^{1/2}XB^{1/2} = f(\frac{i}{2}) = \int_{-\infty}^{\infty} f(t)d\mu_{1/2}(t) + \int_{-\infty}^{\infty} f(i+t)d\nu_{1/2}(t)$$

$$= \int_{-\infty}^{\infty} A^{it}(AX + XB)B^{-it}\frac{dt}{2\cosh(\pi t)}.$$

The unitary invariance of $\|\cdot\|$ thus implies

$$\|A^{1/2}XB^{1/2}\| \leq \|AX + XB\| \int_{-\infty}^{\infty} \frac{dt}{2\cosh(\pi t)}$$

$$= \frac{1}{2}\|AX + XB\|.$$

This completes the proof.   □

Next we will give a generalization of the inequality (4.33). We need some preliminary results.

If a matrix $X = (x_{ij})$ is positive semidefinite, then for any matrix Y we have [52, p.343]

$$\|X \circ Y\| \leq \max_i x_{ii}\|Y\| \tag{4.38}$$

for every unitarily invariant norm.

A function $f : \mathbb{R} \to \mathbb{C}$ is said to be *positive definite* if the matrix $[f(x_i - x_j)] \in M_n$ is positive semidefinite for all choices of points $\{x_1, x_2, \ldots, x_n\} \subset \mathbb{R}$ and all $n = 1, 2, \ldots$.

An integral kernel $K(x, y)$ is said to be *positive definite* on a real interval Δ if

$$\int_\Delta \int_\Delta K(x, y)\bar{f}(x)f(y)dxdy \geq 0$$

for all continuous complex-valued functions f on Δ. It is known [51, p.462] that a continuous kernel $K(x, y)$ is positive definite on Δ if and only if the matrix $[K(x_i, x_j)] \in M_n$ is positive semidefinite for all choices of points $\{x_1, x_2, \ldots, x_n\} \subset \Delta$ and all $n = 1, 2, \ldots$.

Let f be a function in $L^1(\mathbb{R})$. The Fourier transform of f is the function $\hat{f}$ defined as

$$\hat{f}(\xi) = \int_{-\infty}^{\infty} f(x)e^{-i\xi x}dx.$$

When writing Fourier transforms, we ignore constant factors, since the only property of f we use is that of being nonnegative almost everywhere. A well-known theorem of Bochner (see [45, p.70] or [31, p.184]) asserts that $\hat{f}(\xi)$ is positive definite if and only if $f(x) \geq 0$ for almost all x.

Our argument line is as follows: First use Bochner's theorem to show the positive semidefiniteness of a matrix with a special structure; then use the fact (4.38) to obtain a norm inequality.

Lemma 4.22 *Let $\sigma_1, \ldots, \sigma_n$ be positive real numbers. Then for $-1 \leq r \leq 1$ the $n \times n$ matrix*

$$Z = \left(\frac{\sigma_i^r + \sigma_j^r}{\sigma_i + \sigma_j}\right)$$

is positive semidefinite.

Proof. By continuity, it suffices to consider the case $-1 < r < 1$. Since $\sigma_i > 0$, we can put $\sigma_i = e^{x_i}$ for some real x_i. Thus to show that the matrix Z is positive semidefinite it suffices to show that the kernel

$$K(x, y) = \frac{e^{rx} + e^{ry}}{e^x + e^y}$$

is positive definite. Note that

$$K(x, y) = \frac{e^{rx/2}}{e^{x/2}} \left(\frac{e^{r(x-y)/2} + e^{r(y-x)/2}}{e^{(x-y)/2} + e^{(y-x)/2}} \right) \frac{e^{ry/2}}{e^{y/2}}$$

$$= \frac{e^{rx/2}}{e^{x/2}} \left(\frac{\cosh r(x - y)/2}{\cosh(x - y)/2} \right) \frac{e^{ry/2}}{e^{y/2}}.$$

So $K(x, y)$ is positive definite if and only if the kernel

$$L(x, y) = \frac{\cosh r(x - y)/2}{\cosh(x - y)/2}$$

is positive definite, which will follow after we show that

$$f(x) = \frac{\cosh rx}{\cosh x}$$

is a positive definite function on $\mathbb{R}$. The function f is even. Its inverse Fourier transform (see [41, p.1192]) is

$$\check{f}(w) = \frac{\cos(r\pi/2) \cosh(w\pi/2)}{\cos(r\pi) + \cosh(w\pi)}.$$

For $-1 < r < 1$, $\cos(r\pi/2)$ is positive. For all $w \neq 0$, $\cosh(w\pi) > 1$. Thus $\check{f}(w) \geq 0$. By Bochner's theorem, $f(x)$ is positive definite. □

Note that the special case $r = 0$ of Lemma 4.22 is the well-known fact that the Cauchy matrix

$$\left(\frac{1}{\sigma_i + \sigma_j} \right)$$

is positive semidefinite.

Recall that the Schur product theorem says that the Hadamard product of two positive semidefinite matrices is positive semidefinite. Therefore, if $A = (a_{ij})$ is positive semidefinite and $x_1, \ldots, x_n$ are real numbers, then the matrices

$$(a_{ij}^k) \quad \text{and} \quad (x_i x_j a_{ij}) = \text{diag}(x_1, \ldots, x_n) A \text{diag}(x_1, \ldots, x_n)$$

are positive semidefinite for any positive integer k.

Lemma 4.23 *Let $\sigma_1, \ldots, \sigma_n$ be positive numbers, $-1 \leq r \leq 1$, and $-2 < t \leq 2$. Then the $n \times n$ matrix*

$$W = \left(\frac{\sigma_i^r + \sigma_j^r}{\sigma_i^2 + t\sigma_i\sigma_j + \sigma_j^2} \right)$$

is positive semidefinite.

Proof. Let $W = (w_{ij})$. Applying Lemma 4.22 and the Schur product theorem to the formula

$$w_{ij} = \frac{1}{\sigma_i + \sigma_j} \frac{\sigma_i^r + \sigma_j^r}{\sigma_i + \sigma_j} \sum_{k=0}^{\infty} (2-t)^k \left[\frac{\sigma_i\sigma_j}{(\sigma_i + \sigma_j)^2} \right]^k$$

completes the proof.   □

Theorem 4.24 *Let $A, B, X \in M_n$ with A, B positive semidefinite. Then*

$$(2+t)\|A^r X B^{2-r} + A^{2-r} X B^r\| \le 2\|A^2 X + tAXB + XB^2\| \qquad (4.39)$$

for any real numbers r, t satisfying $1 \le 2r \le 3$, $-2 < t \le 2$ and every unitarily invariant norm.

Proof. We first prove the special case $A = B$, i.e.,

$$(2+t)\|A^r X A^{2-r} + A^{2-r} X A^r\| \le 2\|A^2 X + tAXA + XA^2\|. \qquad (4.40)$$

By continuity, without loss of generality, assume that A is positive definite. Let $A = U\Sigma U^*$ be the spectral decomposition with U unitary and $\Sigma = \mathrm{diag}(\sigma_1, \sigma_2, \ldots, \sigma_n)$, each σ_j being positive. Since $\|\cdot\|$ is unitarily invariant, (4.40) is equivalent to

$$(2+t)\|\Sigma^r Q \Sigma^{2-r} + \Sigma^{2-r} Q \Sigma^r\| \le 2\|\Sigma^2 Q + t\Sigma Q \Sigma + Q \Sigma^2\|$$

where $Q = U^* X U$, which may be rewritten as

$$(2+t)\|[(\sigma_i^r \sigma_j^{2-r} + \sigma_i^{2-r}\sigma_j^r)y_{ij}]\| \le 2\|[(\sigma_i^2 + t\sigma_i\sigma_j + \sigma_j^2)y_{ij}]\|$$

for all $Y = (y_{ij}) \in M_n$. This is the same as

$$\|G \circ Z\| \le \|Z\| \quad \text{for all } Z \in M_n \qquad (4.41)$$

where

$$G = \frac{2+t}{2} \left(\frac{\sigma_i^r \sigma_j^{2-r} + \sigma_i^{2-r}\sigma_j^r}{\sigma_i^2 + t\sigma_i\sigma_j + \sigma_j^2} \right) \in M_n.$$

Since $1 \le 2r \le 3$, $-1 \le 2(1-r) \le 1$. By Lemma 4.23,

$$G = \frac{2+t}{2} \Sigma^r \left(\frac{\sigma_i^{2(1-r)} + \sigma_j^{2(1-r)}}{\sigma_i^2 + t\sigma_i\sigma_j + \sigma_j^2} \right) \Sigma^r$$

is positive semidefinite. In fact, G is a correlation matrix; all its diagonal entries are 1. Thus by (4.38), the inequality (4.41) is true, and hence its equivalent form (4.40) is proved.

For the general case, applying (4.40) with A and X replaced by

$$\begin{bmatrix} A & 0 \\ 0 & B \end{bmatrix} \quad \text{and} \quad \begin{bmatrix} 0 & X \\ 0 & 0 \end{bmatrix}$$

respectively completes the proof.   □

Note that the inequality (4.33) corresponds to the case $r = 1, t = 0$ of the inequality (4.39).

Finally we give one more arithmetic-geometric mean inequality.

Theorem 4.25 *Let $A, B \in M_n$ be positive semidefinite. Then*

$$4\|AB\| \le \|(A+B)^2\| \tag{4.42}$$

for every unitarily invariant norm.

Proof. By Theorem 4.19,

$$\|AB\| = \|A^{1/2}(A^{1/2}B^{1/2})B^{1/2}\| \le \frac{1}{2}\|A^{3/2}B^{1/2} + A^{1/2}B^{3/2}\|.$$

So, for (4.42) it suffices to prove

$$2\|A^{3/2}B^{1/2} + A^{1/2}B^{3/2}\| \le \|(A+B)^2\|.$$

We will show more by proving the following singular value inequality for each $1 \le j \le n$:

$$2s_j(A^{3/2}B^{1/2} + A^{1/2}B^{3/2}) \le s_j(A+B)^2. \tag{4.43}$$

Let

$$X = \begin{bmatrix} A^{1/2} & 0 \\ B^{1/2} & 0 \end{bmatrix} \quad \text{and} \quad T = XX^* = \begin{bmatrix} A & A^{1/2}B^{1/2} \\ B^{1/2}A^{1/2} & B \end{bmatrix}.$$

Then T is unitarily equivalent to the matrix

$$G \equiv X^*X = \begin{bmatrix} A+B & 0 \\ 0 & 0 \end{bmatrix}$$

and further

$$T^2 = \begin{bmatrix} \star & A^{3/2}B^{1/2} + A^{1/2}B^{3/2} \\ B^{1/2}A^{3/2} + B^{3/2}A^{1/2} & \star \end{bmatrix}$$

is unitarily equivalent to

$$G^2 = \begin{bmatrix} (A+B)^2 & 0 \\ 0 & 0 \end{bmatrix}.$$

In particular, T^2 and G^2 have the same eigenvalues.

Let $U = I \oplus -I$. Using Theorem 3.14 we have

$$2s_j \begin{bmatrix} A^{3/2}B^{1/2} + A^{1/2}B^{3/2} & 0 \\ 0 & A^{3/2}B^{1/2} + A^{1/2}B^{3/2} \end{bmatrix}$$

$$= 2s_j \begin{bmatrix} 0 & A^{3/2}B^{1/2} + A^{1/2}B^{3/2} \\ B^{1/2}A^{3/2} + B^{3/2}A^{1/2} & 0 \end{bmatrix}$$

$$= s_j(T^2 - UT^2U^*)$$

$$\leq s_j \begin{bmatrix} T^2 & 0 \\ 0 & UT^2U^* \end{bmatrix}$$

$$= s_j \begin{bmatrix} G^2 & 0 \\ 0 & G^2 \end{bmatrix}$$

$$= s_j \begin{bmatrix} (A+B)^2 \oplus 0 & 0 \\ 0 & (A+B)^2 \oplus 0 \end{bmatrix}.$$

Thus

$$2s_j \begin{bmatrix} A^{3/2}B^{1/2} + A^{1/2}B^{3/2} & 0 \\ 0 & A^{3/2}B^{1/2} + A^{1/2}B^{3/2} \end{bmatrix}$$

$$\leq s_j \begin{bmatrix} (A+B)^2 \oplus 0 & 0 \\ 0 & (A+B)^2 \oplus 0 \end{bmatrix}$$

which is the same as (4.43). This completes the proof.   □

Notes and References. For Hilbert space operators, the operator norm version of Theorem 4.19 was first noticed by A. McIntosh [74]. The unitarily invariant norm version of this theorem is due to R. Bhatia and C. Davis [19]; its first proof given here is taken from Bhatia's book [17] and the second proof is due to H. Kosaki [64]. For other proofs of this result see R. A. Horn [50], F. Kittaneh [61] and R. Mathias [73].

The key Lemma 4.23 can be deduced from M. K. Kwong's result [66] on matrix equations (see [87, Lemma 5]). The natural proof using Bochner's theorem given here is due to R. Bhatia and K. R. Parthasarathy [26]. Theorem 4.24 is proved in X. Zhan [87].

Theorem 4.25 is due to R. Bhatia and F. Kittaneh [24].

See F. Hiai and H. Kosaki's two papers [47, 48] for more inequalities on many other means of matrices and Hilbert space operators. See [26] and [64] for more norm inequalities. All these four papers use analytic methods.

4.4 Inequalities of Hölder and Minkowski Types

In this section we derive various matrix versions of the classical Hölder (Cauchy-Schwarz) and Minkowski inequalities.

Lemma 4.26 *Let $X, Y, Z \in M_n$. If*

$$\begin{bmatrix} X & Z \\ Z^* & Y \end{bmatrix} \geq 0$$

then

$$\| \, |Z|^r \, \| \leq \|X^{pr/2}\|^{1/p} \|Y^{qr/2}\|^{1/q}$$

for all positive numbers r, p, q with $p^{-1} + q^{-1} = 1$ and every unitarily invariant norm.

Proof. By Lemma 1.21, there exists a contraction G such that $Z = X^{1/2} G Y^{1/2}$. Let $\gamma = (\gamma_1, \ldots, \gamma_n)$ with $\gamma_1 \geq \cdots \geq \gamma_n \geq 0$. Using Theorem 2.5 we have

$$\{\gamma_j s_j^r(Z)\} \prec_{wlog} \{\gamma_j^{1/p} s_j^{r/2}(X) \cdot \gamma_j^{1/q} s_j^{r/2}(Y)\}.$$

Since weak log-majorization implies weak majorization (Theorem 2.7), we get

$$\{\gamma_j s_j^r(Z)\} \prec_w \{\gamma_j^{1/p} s_j^{r/2}(X) \cdot \gamma_j^{1/q} s_j^{r/2}(Y)\}$$

from which follows via the classical Hölder inequality

$$\sum_{j=1}^n \gamma_j s_j^r(Z) \leq \left[\sum_{j=1}^n \gamma_j s_j^{pr/2}(X)\right]^{1/p} \left[\sum_{j=1}^n \gamma_j s_j^{qr/2}(Y)\right]^{1/q},$$

i.e.,

$$\| \, |Z|^r \, \|_\gamma \leq \|X^{pr/2}\|_\gamma^{1/p} \|Y^{qr/2}\|_\gamma^{1/q}. \tag{4.44}$$

See the paragraph preceding Lemma 4.1 for the notation $\| \cdot \|_\gamma$. By Lemma 4.1, given any unitarily invariant norm $\| \cdot \|$ there exists a set K depending only on $\| \cdot \|$ such that

$$\|A\| = \max_{\gamma \in K} \|A\|_\gamma \quad \text{for all } A \in M_n.$$

Thus, taking maximum on both sides of (4.44) completes the proof. □

Setting $X = AA^*$, $Y = B^*B$ and $Z = AB$ in Lemma 4.26 gives the following

Corollary 4.27 *For all $A, B \in M_n$ and every unitarily invariant norm,*

$$\| \, |AB|^r \, \| \leq \| \, |A|^{pr} \, \|^{1/p} \| \, |B|^{qr} \, \|^{1/q} \tag{4.45}$$

where r, p, q are positive real numbers with $p^{-1} + q^{-1} = 1$.

The next lemma is a consequence of Theorem 2.9.

Lemma 4.28 *Let A, B be positive semidefinite matrices. Then for real numbers $r > 0$, $0 < s < t$,*

$$\{\lambda_j^{r/s}(A^s B^s)\} \prec_w \{\lambda_j^{r/t}(A^t B^t)\}.$$

The following result is a matrix Hölder inequality

Theorem 4.29 *Let $A, B, X \in M_n$ with A, B positive semidefinite. Then*

$$\| |AXB|^r \| \le \| |A^p X|^r \|^{1/p} \| |XB^q|^r \|^{1/q} \qquad (4.46)$$

for all positive real numbers r, p, q with $p^{-1} + q^{-1} = 1$ and every unitarily invariant norm.

Proof. Let $X = UT$ be the polar decomposition with U unitary and $T = |X|$. Write $AXB = (AUT^{1/p})(T^{1/q}B)$ and use (4.45) to obtain

$$\||AXB|^r\| \le \|(T^{1/p}U^* A^2 UT^{1/p})^{pr/2}\|^{1/p}\|(BT^{2/q}B)^{qr/2}\|^{1/q}. \qquad (4.47)$$

Since YZ and ZY have the same eigenvalues, Lemma 4.28 ensures that

$$\begin{aligned}
\{\lambda_j^{pr/2}(T^{1/p}U^* A^2 UT^{1/p})\} &= \{\lambda_j^{pr/2}[(A^{2p})^{1/p}(UT^2 U^*)^{1/p}]\} \\
&\prec_w \{\lambda_j^{r/2}(A^{2p}UT^2 U^*)\} \quad \text{(since } p^{-1} < 1\text{)} \\
&= \{\lambda_j^{r/2}(A^{2p}XX^*)\} \\
&= \{\lambda_j^{r/2}[(A^p X)^*(A^p X)]\} \\
&= \{s_j^r(A^p X)\}
\end{aligned}$$

and

$$\begin{aligned}
\{\lambda_j^{qr/2}(BT^{2/q}B)\} &= \{\lambda_j^{qr/2}[(T^2)^{1/q}(B^{2q})^{1/q}]\} \\
&\prec_w \{\lambda_j^{r/2}(T^2 B^{2q})\} \quad \text{(since } q^{-1} < 1\text{)} \\
&= \{\lambda_j^{r/2}(X^* X B^{2q})\} \\
&= \{\lambda_j^{r/2}[(XB^q)^*(XB^q)]\} \\
&= \{s_j^r(XB^q)\}.
\end{aligned}$$

Thus

$$\{\lambda_j^{pr/2}(T^{1/p}U^* A^2 UT^{1/p})\} \prec_w \{s_j^r(A^p X)\} \qquad (4.48)$$

and

$$\{\lambda_j^{qr/2}(BT^{2/q}B)\} \prec_w \{s_j^r(XB^q)\}. \qquad (4.49)$$

By Fan's dominance principle, the weak majorizations (4.48) and (4.49) imply

$$\|(T^{1/p}U^*A^2UT^{1/p})^{pr/2}\| \leq \||A^pX|^r\|$$

and

$$\|(BT^{2/q}B)^{qr/2}\| \leq \||XB^q|^r\|$$

for every unitarily invariant norm. Combining these two inequalities and (4.47) gives (4.46). □

Now let us consider some special cases of Theorem 4.29.

The case $p = q = 2$ of (4.46) is the following matrix Cauchy-Schwarz inequality:

$$\||A^{1/2}XB^{1/2}|^r\|^2 \leq \||AX|^r\| \cdot \||XB|^r\| \tag{4.50}$$

for $A, B \geq 0$, any X and $r > 0$, which is equivalent to

$$\||AXB^*|^r\|^2 \leq \||A^*AX|^r\| \cdot \||XB^*B|^r\| \tag{4.51}$$

for all A, B, X and $r > 0$.

The special case $X = I$ of (4.51) has the following simple form:

$$\||A^*B|^r\|^2 \leq \|(A^*A)^r\| \cdot \|(B^*B)^r\| \tag{4.52}$$

The case $r = 1$ of (4.46) is the following inequality:

$$\|AXB\| \leq \|A^pX\|^{1/p}\|XB^q\|^{1/q}. \tag{4.53}$$

Again we can use the integral formula (4.36) with the same function f as in the second proof of Theorem 4.19 and $\theta = 1/q$ to give a direct proof of (4.53) as follows.

$$A^{1/p}XB^{1/q} = f(\frac{i}{q}) = \int_{-\infty}^{\infty} f(t)d\mu_{1/q}(t) + \int_{-\infty}^{\infty} f(i+t)d\nu_{1/q}(t)$$
$$= \int_{-\infty}^{\infty} A^{it}(AX)B^{-it}d\mu_{1/q}(t) + \int_{-\infty}^{\infty} A^{it}(XB)B^{-it}d\nu_{1/q}(t).$$

Hence

$$\|A^{1/p}XB^{1/q}\| \leq \frac{\|AX\|}{p} + \frac{\|XB\|}{q}$$

for every unitarily invariant norm. Then for any real number $\alpha > 0$, replacing A, B by $\alpha^p A$, $\alpha^{-q}B$ respectively we get

$$\|A^{1/p}XB^{1/q}\| \leq \frac{\alpha^p}{p}\|AX\| + \frac{\alpha^{-q}}{q}\|XB\|.$$

Since the minimum of the right-hand side over all $\alpha > 0$ is $\|AX\|^{1/p}\|XB\|^{1/q}$, we obtain (4.53).

Now we refine the Cauchy-Schwarz inequality (4.50).

Theorem 4.30 *Let $A, B, X \in M_n$ with A, B positive semidefinite and X arbitrary. For any positive real number r and every unitarily invariant norm, the function*

$$\phi(t) = \| |A^t X B^{1-t}|^r \| \cdot \| |A^{1-t} X B^t|^r \|$$

is convex on the interval $[0, 1]$ and attains its minimum at $t = 1/2$. Consequently, it is decreasing on $[0, 1/2]$ and increasing on $[1/2, 1]$.

Proof. Since $\phi(t)$ is continuous and symmetric with respect to $t = 1/2$, all the conclusions will follow after we show that

$$\phi(t) \leq \{\phi(t + s) + \phi(t - s)\}/2 \tag{4.54}$$

for $t \pm s \in [0, 1]$. By (4.50) we have

$$\| |A^t X B^{1-t}|^r \| = \| |A^s (A^{t-s} X B^{1-t-s}) B^s|^r \|$$
$$\leq \{\| |A^{t+s} X B^{1-(t+s)}|^r \| \cdot \| |A^{t-s} X B^{1-(t-s)}|^r \|\}^{1/2}$$

and

$$\| |A^{1-t} X B^t|^r \| = \| |A^s (A^{1-t-s} X B^{t-s}) B^s|^r \|$$
$$\leq \{\| |A^{1-(t-s)} X B^{t-s}|^r \| \cdot \| |A^{1-(t+s)} X B^{t+s}|^r \|\}^{1/2}.$$

Upon multiplication of the above two inequalities we obtain

$$\| |A^t X B^{1-t}|^r \| \cdot \| |A^{1-t} X B^t|^r \| \leq \{\phi(t + s)\phi(t - s)\}^{1/2}. \tag{4.55}$$

Applying the arithmetic-geometric mean inequality to the right-hand side of (4.55) yields (4.54). This completes the proof. $\square$

An immediate consequence of Theorem 4.30 interpolates the Cauchy-Schwarz inequality (4.50) as follows.

Corollary 4.31 *Let $A, B, X \in M_n$ with A, B positive semidefinite and X arbitrary. For any $r > 0$ and every unitarily invariant norm,*

$$\| |A^{1/2} X B^{1/2}|^r \|^2 \leq \| |A^t X B^{1-t}|^r \| \cdot \| |A^{1-t} X B^t|^r \|$$
$$\leq \| |AX|^r \| \cdot \| |XB|^r \|$$

holds for $0 \leq t \leq 1$.

Another consequence of Theorem 4.30 is the following corollary.

Corollary 4.32 *Let $A, B, X \in M_n$ with A, B positive definite and X arbitrary. For any $r > 0$ and every unitarily invariant norm, the function*

$$\psi(s) = \| |A^s X B^s|^r \| \cdot \| |A^{-s} X B^{-s}|^r \|$$

is convex on $(-\infty, \infty)$, attains its minimum at $s = 0$, and hence it is decreasing on $(-\infty, 0)$ and increasing on $(0, \infty)$.

Proof. In Theorem 4.30, replacing A, B, X and t by $A^2, B^{-2}, A^{-1}XB$ and $(1+s)/2$ respectively, we see that $\psi(s)$ is convex on $(-1, 1)$, decreasing on $(-1, 0)$, increasing on $(0, 1)$ and attains its minimum at $s = 0$ when $-1 \leq s \leq 1$. Next replacing A, B by their appropriate powers it is easily seen that the above convexity and monotonicity of $\psi(s)$ on those intervals are equivalent to the same properties on $(-\infty, \infty)$, $(-\infty, 0)$ and $(0, \infty)$ respectively.   □

Given a norm $\| \cdot \|$ on M_n, the condition number of an invertible matrix A is defined as

$$c(A) = \|A\| \cdot \|A^{-1}\|.$$

This is one of the basic concepts in numerical analysis; it serves as measures of the difficulty in solving a system of linear equations.

The special case $r = 1$, $X = B = I$ of Corollary 4.32 gives the following interesting result.

Corollary 4.33 *Let A be positive definite. Then for every unitarily invariant norm,*

$$c(A^s) = \|A^s\| \cdot \|A^{-s}\|$$

is increasing in $s > 0$.

Due to Theorems 4.15 and 4.16, $\| |X|^p \|^{1/p}$ for $p = \infty$ is understood as $\|X\|_\infty$ and $\| |A|^p + |B|^p \|^{1/p}$ for $p = \infty$ as $\| |A| \vee |B| \|_\infty$.

It is easy to see that the special case $X = I$ of (4.53) can be written equivalently as

$$\|Y^*X\| \leq \| |X|^p \|^{1/p} \| |Y|^q \|^{1/q} \tag{4.56}$$

for all X and Y. Here note that $\| |X^*|^r \| = \| |X|^r \|$ for any $r > 0$ and every unitarily invariant norm.

We will use the following fact: Let A and B be positive semidefinite matrices having the eigenvalues $\alpha_1 \geq \cdots \geq \alpha_n \, (\geq 0)$ and $\beta_1 \geq \cdots \geq \beta_n \, (\geq 0)$, respectively. If $\alpha_i \leq \beta_i$ $(i = 1, \ldots, n)$ (in particular, if $A \leq B$), then there exists a unitary U such that $A^r \leq UB^rU^*$ for all $r > 0$.

The next result is another matrix Hölder inequality.

Theorem 4.34 *Let $1 \leq p, q \leq \infty$ with $p^{-1} + q^{-1} = 1$. Then for all $A, B, C, D \in M_n$ and every unitarily invariant norm,*

$$2^{-|\frac{1}{p}-\frac{1}{2}|}\|C^*A + D^*B\| \leq \| |A|^p + |B|^p \|^{1/p} \| |C|^q + |D|^q \|^{1/q}. \tag{4.57}$$

Moreover, the constant $2^{-|\frac{1}{p}-\frac{1}{2}|}$ is best possible.

Proof. Since $|\frac{1}{p} - \frac{1}{2}| = |\frac{1}{q} - \frac{1}{2}|$ and the inequality is symmetric with respect to p and q, we may assume $1 \leq p \leq 2 \leq q \leq \infty$. Note that

$$\begin{bmatrix} C & 0 \\ D & 0 \end{bmatrix}^* \begin{bmatrix} A & 0 \\ B & 0 \end{bmatrix} = \begin{bmatrix} C^*A + D^*B & 0 \\ 0 & 0 \end{bmatrix}.$$

From (4.56) it follows that

$$\|C^*A + D^*B\| = \left\| \begin{bmatrix} C & 0 \\ D & 0 \end{bmatrix}^* \begin{bmatrix} A & 0 \\ B & 0 \end{bmatrix} \right\|$$

$$\leq \left\| \left| \begin{bmatrix} A & 0 \\ B & 0 \end{bmatrix} \right|^p \right\|^{1/p} \cdot \left\| \left| \begin{bmatrix} C & 0 \\ D & 0 \end{bmatrix} \right|^q \right\|^{1/q}$$

$$= \|(|A|^2 + |B|^2)^{p/2}\|^{1/p} \|(|C|^2 + |D|^2)^{q/2}\|^{1/q}.$$

Since $1 \leq p \leq 2$, (4.11) implies

$$\|(|A|^2 + |B|^2)^{p/2}\| \leq \| |A|^p + |B|^p \|.$$

Since the operator concavity of $t^{2/q}$ gives

$$\frac{|C|^2 + |D|^2}{2} \leq \left(\frac{|C|^q + |D|^q}{2} \right)^{2/q},$$

by the remark preceding the theorem we get

$$\left(\frac{|C|^2 + |D|^2}{2} \right)^{q/2} \leq U \left(\frac{|C|^q + |D|^q}{2} \right) U^*$$

for some unitary U. Therefore we have

$$\|(|C|^2 + |D|^2)^{q/2}\|^{1/q} \leq 2^{\frac{1}{2} - \frac{1}{q}} \| |C|^q + |D|^q \|^{1/q}$$
$$= 2^{\frac{1}{p} - \frac{1}{2}} \| |C|^q + |D|^q \|^{1/q}.$$

Thus the desired inequality (4.57) follows.

The best possibility of the constant is seen from the following example:

$$A = C = D = \begin{bmatrix} 1 & 0 \\ 0 & 0 \end{bmatrix}, \quad B = \begin{bmatrix} 0 & 1 \\ 0 & 0 \end{bmatrix}$$

with the operator norm $\| \cdot \|_\infty$. $\square$

In particular, the case $p = q = 2$ of (4.57) is

$$\|C^*A + D^*B\|^2 \leq \| |A|^2 + |B|^2 \| \cdot \| |C|^2 + |D|^2 \|.$$

The following result is a matrix Minkowski inequality.

Theorem 4.35 *Let $1 \leq p < \infty$. For $A_i, B_i \in M_n$ $(i = 1, 2)$ and every unitarily invariant norm,*

$$2^{-|\frac{1}{p} - \frac{1}{2}|} \| |A_1 + A_2|^p + |B_1 + B_2|^p \|^{1/p}$$
$$\leq \| |A_1|^p + |B_1|^p \|^{1/p} + \| |A_2|^p + |B_2|^p \|^{1/p}.$$

Proof. Since

$$\|(|A|^2 + |B|^2)^{p/2}\|^{1/p} = \left\|\left|\begin{bmatrix} A & 0 \\ B & 0 \end{bmatrix}\right|^p\right\|^{1/p}$$

is a norm in (A, B), we have

$$\|(|A_1 + A_2|^2 + |B_1 + B_2|^2)^{p/2}\|^{1/p}$$
$$\leq \|(|A_1|^2 + |B_1|^2)^{p/2}\|^{1/p} + \|(|A_2|^2 + |B_2|^2)^{p/2}\|^{1/p}. \qquad (4.58)$$

When $1 \leq p \leq 2$, (4.11) implies

$$\|(|A_i|^2 + |B_i|^2)^{p/2}\| \leq \||A_i|^p + |B_i|^p\| \qquad (i = 1, 2). \qquad (4.59)$$

By the operator concavity of $t \mapsto t^{p/2}$ we get

$$\frac{|A_1 + A_2|^p + |B_1 + B_2|^p}{2} \leq \left(\frac{|A_1 + A_2|^2 + |B_1 + B_2|^2}{2}\right)^{p/2},$$

so that

$$2^{\frac{p}{2}-1}\||A_1 + A_2|^p + |B_1 + B_2|^p\| \leq \|(|A_1 + A_2|^2 + |B_1 + B_2|^2)^{p/2}\|. \qquad (4.60)$$

Combining (4.60), (4.58) and (4.59) we have

$$2^{\frac{1}{2}-\frac{1}{p}}\||A_1 + A_2|^p + |B_1 + B_2|^p\|^{1/p}$$
$$\leq \||A_1|^p + |B_1|^p\|^{1/p} + \||A_2|^p + |B_2|^p\|^{1/p}.$$

When $p \geq 2$, (4.12) implies

$$\||A_1 + A_2|^p + |B_1 + B_2|^p\| \leq \|(|A_1 + A_2|^2 + |B_1 + B_2|^2)^{p/2}\|. \qquad (4.61)$$

Since, as in the proof of Theorem 4.34,

$$\left(\frac{|A_i|^2 + |B_i|^2}{2}\right)^{p/2} \leq U_i \left(\frac{|A_i|^p + |B_i|^p}{2}\right) U_i^*$$

for some unitary U_i, we have

$$2^{1-\frac{p}{2}}\|(|A_i|^2 + |B_i|^2)^{p/2}\| \leq \||A_i|^p + |B_i|^p\| \qquad (i = 1, 2). \qquad (4.62)$$

Combining (4.61), (4.58) and (4.62) yields

$$2^{\frac{1}{p}-\frac{1}{2}}\||A_1 + A_2|^p + |B_1 + B_2|^p\|^{1/p}$$
$$\leq \||A_1|^p + |B_1|^p\|^{1/p} + \||A_2|^p + |B_2|^p\|^{1/p}.$$

This completes the proof. □

The example

$$A_1 = \begin{bmatrix} 1 & 0 \\ 0 & 0 \end{bmatrix}, \quad A_2 = \begin{bmatrix} 0 & 1 \\ 0 & 0 \end{bmatrix}, \quad B_1 = \begin{bmatrix} 0 & 0 \\ 0 & 1 \end{bmatrix}, \quad B_2 = \begin{bmatrix} 0 & 0 \\ 1 & 0 \end{bmatrix}$$

with the spectral norm shows that when $1 \le p \le 2$ is fixed and $\|\cdot\|$ is arbitrary, the constant $2^{\frac{1}{2}-\frac{1}{p}}$ in Theorem 4.35 is best possible.

When A_i, B_i $(i = 1, 2)$ are positive semidefinite matrices, there is a possibility to obtain a sharper inequality. When $\|\cdot\| = \|\cdot\|_\infty$ and $1 \le p \le 2$, it is proved in [10, Proposition 3.7] that

$$\|(A_1 + A_2)^p + (B_1 + B_2)^p\|_\infty^{1/p} \le \|A_1^p + B_1^p\|_\infty^{1/p} + \|A_2^p + B_2^p\|_\infty^{1/p}.$$

We also have

$$2^{\frac{1}{p}-1}\|(A_1 + A_2)^p + (B_1 + B_2)^p\|^{1/p} \le \|A_1^p + B_1^p\|^{1/p} + \|A_2^p + B_2^p\|^{1/p}$$

for every unitarily invariant norm and $1 \le p \le 2$. (The constant $2^{\frac{1}{p}-1}$ is better than $2^{-|\frac{1}{p}-\frac{1}{2}|}$ for $1 \le p < 4/3$.) Indeed, since the operator convexity of t^p gives

$$2^{1-p}(A_1 + A_2)^p \le A_1^p + A_2^p, \quad 2^{1-p}(B_1 + B_2)^p \le B_1^p + B_2^p,$$

we get

$$2^{\frac{1}{p}-1}\|(A_1 + A_2)^p + (B_1 + B_2)^p\|^{1/p}$$
$$\le \|A_1^p + A_2^p + B_1^p + B_2^p\|^{1/p}$$
$$\le (\|A_1^p + B_1^p\| + \|A_2^p + B_2^p\|)^{1/p}$$
$$\le \|A_1^p + B_1^p\|^{1/p} + \|A_2^p + B_2^p\|^{1/p}.$$

Notes and References. Theorem 4.29 is proved in H. Kosaki [64, Theorem 3] and independently in R. A. Horn and X. Zhan [55, Theorem 3]; the proof given here is taken from [55]. A similar result is proved in F. Hiai [46, p.174].

The inequality (4.51) is due to R. Bhatia and C. Davis [20]. The inequality (4.52) is due to R. A. Horn and R. Mathias [53, 54]. The inequality (4.53) is due to F. Kittaneh [62]; its proof using the integral formula given here is in [64].

Theorem 4.30, Corollaries 4.31 and 4.32, Theorems 4.34 and 4.35 are proved by F. Hiai and X. Zhan [49]. Corollary 4.33 is due to A. W. Marshall and I. Olkin [71].

See [49] for more inequalities of Minkowski type.

4.5 Permutations of Matrix Entries

In this section we study norm variations of matrices under permutations of their entries.

A map $\Phi : M_n \to M_n$ is called a *permutation operator* if for all A the entries of $\Phi(A)$ are one fixed rearrangement of those of A. Familiar examples of permutation operators are the transpose operation, permutations of rows and columns. A basic observation is that every permutation operator is an invertible linear map.

Let M_n be endowed with a norm $\| \cdot \|$ and $\Phi : M_n \to M_n$ be a linear map. We use the same notation for the operator norm of Φ :

$$\|\Phi\| \equiv \sup\{\|\Phi(A)\| : \|A\| \leq 1,\ A \in M_n\}.$$

The following result will simplify our investigations. It says that the maximum of operator norms of a linear map on matrix spaces induced by unitarily invariant norms is attained at the norm induced by either the spectral norm or the trace norm.

Lemma 4.36 *Let* $\Phi : M_n \to M_n$ *be a linear map. Then for every unitarily invariant norm* $\| \cdot \|$ *on* M_n

$$\|\Phi\| \leq \max\{\|\Phi\|_\infty, \|\Phi\|_1\}. \tag{4.63}$$

Proof. Denote by N_{ui} the set of unitarily invariant norms on M_n. By Fan's dominance principle (Lemma 4.2) we have

$$\max\{\|\Phi\| : \| \cdot \| \in N_{ui}\} = \max\{\|\Phi\|_{(k)} : 1 \leq k \leq n\}. \tag{4.64}$$

On the other hand, Lemma 4.17 asserts that for any $T \in M_n$ and each $1 \leq k \leq n$

$$\|T\|_{(k)} = \min\{k\|X\|_\infty + \|Y\|_1 : T = X + Y\}. \tag{4.65}$$

Suppose $\|\Phi\|_{(k)} = \|\Phi(A)\|_{(k)}$ with $\|A\|_{(k)} = 1$. Then by (4.65) there exist $X, Y \in M_n$ such that

$$A = X + Y, \quad 1 = \|A\|_{(k)} = k\|X\|_\infty + \|Y\|_1. \tag{4.66}$$

Note that Φ is linear. Using (4.65) again and (4.66) we get

$$\begin{aligned}
\|\Phi\|_{(k)} = \|\Phi(A)\|_{(k)} &= \|\Phi(X) + \Phi(Y)\|_{(k)} \\
&\leq k\|\Phi(X)\|_\infty + \|\Phi(Y)\|_1 \\
&= k\|X\|_\infty \cdot \frac{\|\Phi(X)\|_\infty}{\|X\|_\infty} + \|Y\|_1 \cdot \frac{\|\Phi(Y)\|_1}{\|Y\|_1} \\
&\leq \max\{\frac{\|\Phi(X)\|_\infty}{\|X\|_\infty}, \frac{\|\Phi(Y)\|_1}{\|Y\|_1}\} \\
&\leq \max\{\|\Phi\|_\infty, \|\Phi\|_1\}.
\end{aligned}$$

Thus

$$\|\Phi\|_{(k)} \le \max\{\|\Phi\|_\infty, \|\Phi\|_1\}.$$

Combining the above inequality and (4.64) we obtain (4.63).   □

Given a fixed $A \in M_n$, let us consider the Hadamard multiplier $T_A :$ $M_n \to M_n$ defined as $T_A(X) = A \circ X$. It is easily seen that the adjoint operator of T_A with respect to the Frobenius inner product $\langle X, Y \rangle = \mathrm{tr} XY^*$ is $T_A^* = T_{\bar{A}}$ ($\bar{A}$ denotes the complex conjugate matrix of A). On the other hand, the spectral norm and the trace norm are dual to each other in the sense that

$$\|A\|_\infty = \sup_{X \neq 0} \frac{|\langle A, X \rangle|}{\|X\|_1} \quad \text{and} \quad \|A\|_1 = \sup_{X \neq 0} \frac{|\langle A, X \rangle|}{\|X\|_\infty}.$$

Note also that $\|T_{\bar{A}}\|_\infty = \|T_A\|_\infty$. Therefore the following corollary is a consequence of Lemma 4.36.

Corollary 4.37 *Given $A \in M_n$, let T_A be the Hadamard multiplier induced by A. Then for any unitarily invariant norm $\|\cdot\|$*

$$\|T_A\| \le \|T_A\|_\infty.$$

This corollary indicates that for some unitarily invariant norm problems involving Hadamard products, it suffices to consider the spectral norm case.

The Frobenius norm $\|(a_{ij})\|_2 = (\sum_{i,j} |a_{ij}|^2)^{1/2}$ plays the role of a bridge in our analysis. Recall that $\|A\|_2 = (\sum_j s_j(A)^2)^{1/2}$.

Theorem 4.38 *For every permutation operator Φ and all unitarily invariant norms $\|\cdot\|$ on M_n,*

$$\frac{1}{\sqrt{n}}\|A\| \le \|\Phi(A)\| \le \sqrt{n}\|A\|, \quad A \in M_n \tag{4.67}$$

and the constants $\sqrt{n}$ and $1/\sqrt{n}$ are best possible.

Proof. The first inequality follows from the second. Since a permutation operator is bijective, in the second inequality we may replace A by $\Phi^{-1}(A)$ and then replace Φ by Φ^{-1}. Thus it suffices to prove the second inequality. The conclusion is equivalent to $\|\Phi\| \le \sqrt{n}$. By Lemma 4.36 it suffices to show $\|\Phi\|_\infty \le \sqrt{n}$ and $\|\Phi\|_1 \le \sqrt{n}$. But this follows from

$$\|\Phi(A)\|_\infty \le \|\Phi(A)\|_2 = \|A\|_2 \le \sqrt{n}\|A\|_\infty$$

and

$$\|\Phi(A)\|_1 \le \sqrt{n}\|\Phi(A)\|_2 = \sqrt{n}\|A\|_2 \le \sqrt{n}\|A\|_1.$$

Now consider the permutation operator Ψ which interchanges the first column and the diagonal of a matrix and keeps all other entries fixed. Let $I \in M_n$ be the identity matrix. Then $\|\Psi(I)\|_\infty = \sqrt{n}\|I\|_\infty = \sqrt{n}$. Let $e \in \mathbb{C}^n$ be the vector with each component 1 and $B = (e, 0, \ldots, 0)$. Then $\|\Psi(B)\|_1 = \sqrt{n}\|B\|_1 = n$. Thus we see that the constant $\sqrt{n}$ in (4.67) is best possible for the operator norm and the trace norm. $\square$

For real numbers $x_1 \geq x_2 \geq \cdots \geq x_n \geq 0$ and integer $1 \leq k \leq n$, we have

$$x_1 + \cdots + x_k \leq \sqrt{k}(x_1^2 + \cdots + x_n^2)^{1/2} \leq \sqrt{n}(x_1 + \cdots + x_k).$$

We can use these two inequalities to give another proof of Theorem 4.38 as follows.

$$\|\Phi(A)\|_{(k)} \leq \sqrt{k}\|\Phi(A)\|_2 = \sqrt{k}\|A\|_2 \leq \sqrt{n}\|A\|_{(k)}.$$

Thus

$$\|\Phi(A)\|_{(k)} \leq \sqrt{n}\|A\|_{(k)}, \quad k = 1, 2, \ldots, n.$$

Finally applying the Fan dominance principle gives (4.67).

For general Schatten p-norms, the estimate (4.67) can be improved.

Theorem 4.39 *For any permutation operator Φ on M_n and any $A \in M_n$,*

$$n^{\frac{1}{p}-\frac{1}{2}}\|A\|_p \leq \|\Phi(A)\|_p \leq n^{\frac{1}{2}-\frac{1}{p}}\|A\|_p, \quad \text{if} \quad 2 \leq p \leq \infty, \tag{4.68}$$

$$n^{\frac{1}{2}-\frac{1}{p}}\|A\|_p \leq \|\Phi(A)\|_p \leq n^{\frac{1}{p}-\frac{1}{2}}\|A\|_p, \quad \text{if} \quad 1 \leq p \leq 2 \tag{4.69}$$

and all these inequalities are sharp.

Proof. Again the left-hand side inequalities follow from the corresponding right-hand side inequalities. Using the fact that if $p \geq 2$ then for any $x \in \mathbb{C}^n$, $\|x\|_2 \leq n^{\frac{1}{2}-\frac{1}{p}}\|x\|_p$ we have

$$\|\Phi(A)\|_p \leq \|\Phi(A)\|_2 = \|A\|_2 \leq n^{\frac{1}{2}-\frac{1}{p}}\|A\|_p.$$

This proves (4.68).

Now assume $1 \leq p \leq 2$. Then for any $x \in \mathbb{C}^n$, $\|x\|_p \leq n^{\frac{1}{p}-\frac{1}{2}}\|x\|_2$. Consequently

$$\|\Phi(A)\|_p \leq n^{\frac{1}{p}-\frac{1}{2}}\|\Phi(A)\|_2 = n^{\frac{1}{p}-\frac{1}{2}}\|A\|_2 \leq n^{\frac{1}{p}-\frac{1}{2}}\|A\|_p.$$

This proves (4.69).

The examples with Ψ, I, B in the proof of Theorem 4.38 also show both (4.68) and (4.69) are sharp. $\square$

The idea in the above proofs is simple: The Frobenius norm is invariant under permutations of matrix entries. The next result shows that the converse is also true.

Theorem 4.40 Let $\| \cdot \|$ be a unitarily invariant norm on M_n. If $\|\Phi(A)\| = \|A\|$ holds for all permutation operators Φ and all $A \in M_n$, then $\| \cdot \|$ is a constant multiple of $\| \cdot \|_2$.

Proof. Let Ψ be as in the proof of Theorem 4.38 and $A = \operatorname{diag}(s_1, s_2, \ldots, s_n)$ where $(s_1, s_2, \ldots, s_n) \in \mathbb{R}_+^n$. Let φ be the symmetric gauge function corresponding to $\| \cdot \|$. Then

$$\varphi(s_1, s_2, \ldots, s_n) = \|A\| = \|\Psi(A)\| = \varphi((\sum_1^n s_j^2)^{1/2}, 0, \ldots, 0)$$

$$= (\sum_1^n s_j^2)^{1/2} \varphi(1, 0, \ldots, 0)$$

$$= \varphi(1, 0, \ldots, 0)\|A\|_2.$$

Since $s_1, s_2, \ldots, s_n$ are arbitrary, $\| \cdot \| = \varphi(1, 0, \ldots, 0)\| \cdot \|_2$. □

A map $\Phi : M_n \to M_n$ is called *positive* if it preserves the set $\mathcal{P}_n$ of positive semidefinite matrices in M_n, i.e., $\Phi(\mathcal{P}_n) \subseteq \mathcal{P}_n$. It is not hard to prove that if $\Phi : M_n \to M_n$ is a positive permutation operator, then there exists a permutation matrix P such that

$$\Phi(A) = PAP^t \quad \text{for all } A \in M_n$$

or

$$\Phi(A) = PA^tP^t \quad \text{for all } A \in M_n$$

where X^t denotes the transpose of X. This result shows that among the permutation operators, only the transpose operation and simultaneous row and column permutations can preserve positive semidefiniteness.

Notes and Reference. The results in this section seem new. See [91] for the forms of those permutation operators that preserve eigenvalues, singular values, etc.

4.6 The Numerical Radius

Let $\| \cdot \|$ be the usual Euclidean norm on $\mathbb{C}^n$. The *numerical range* (or the field of values) of a matrix $A \in M_n$ is defined as

$$W(A) \equiv \{\langle Ax, x \rangle : \|x\| = 1, x \in \mathbb{C}^n\}.$$

For any $A \in M_n$, $W(A)$ is a convex compact subset of the complex plane containing the eigenvalues of A. See [52, Chapter 1] for this topic.

The *numerical radius* of $A \in M_n$ is by definition

$$w(A) \equiv \max\{|z| : z \in W(A)\}.$$

It is easy to see that $w(\cdot)$ is a norm on M_n and we have

$$\frac{1}{2}\|A\|_\infty \leq w(A) \leq \|A\|_\infty.$$

The second inequality is obvious while the first inequality can be shown by the Cartesian decomposition.

We call a norm $\|\cdot\|$ on M_n *weakly unitarily invariant* if

$$\|UAU^*\| = \|A\|$$

for all A and all unitary U. Evidently the numerical radius is such a norm. Since

$$w\left(\begin{bmatrix} 0 & 2 \\ 0 & 0 \end{bmatrix}\right) = 1,$$

the numerical radius is not unitarily invariant (this example in fact shows that $w(\cdot)$ is even not permutation-invariant); it is not submultiplicative:

$$w(AB) \leq w(A)w(B)$$

is in general false even for commuting A, B. Consider the nilpotent matrix

$$A = \begin{bmatrix} 0 & 1 & 0 & 0 \\ 0 & 0 & 1 & 0 \\ 0 & 0 & 0 & 1 \\ 0 & 0 & 0 & 0 \end{bmatrix}$$

and set $B = A^2$. Then $w(A) < 1$, $w(A^2) = 1/2$ and $w(A^3) = 1/2$. But the next best thing is indeed true. We have the following

Theorem 4.41 (Power Inequality) *For every matrix A and any positive integer k,*

$$w(A^k) \leq w(A)^k. \tag{4.70}$$

Proof. Since $w(\cdot)$ is positively homogeneous, it suffices to prove $w(A^k) \leq 1$ under the assumption $w(A) \leq 1$.

Denote by $\mathrm{Re}A = (A + A^*)/2$ the real part of a matrix A. Let a, z be complex numbers. Then $|a| \leq 1$ if and only if $\mathrm{Re}(1 - za) \geq 0$ for each z with $|z| < 1$. Consequently

$$w(A) \leq 1 \text{ if and only if } \mathrm{Re}(I - zA) \geq 0 \text{ for } |z| < 1. \tag{4.71}$$

For a nonsingular matrix B, $B+B^* = B[B^{-1}+(B^{-1})^*]B^*$. Thus $\mathrm{Re}B \geq 0$ if and only if $\mathrm{Re}(B^{-1}) \geq 0$. From (4.71) we get

$$w(A) \leq 1 \text{ if and only if } \mathrm{Re}(I - zA)^{-1} \geq 0 \text{ for } |z| < 1. \tag{4.72}$$

Observe next that if $\omega = e^{2\pi i/k}$, a primitive kth root of unity, then

$$\frac{1}{1-z^k} = \frac{1}{k}\sum_{j=0}^{k-1}\frac{1}{1-\omega^j z}$$

for all z other than the powers of ω. This identity can be verified as follows. First we have the identity $1 - z^k = \prod_{j=0}^{k-1}(1 - \omega^j z)$. Multiply both sides of the identity in question by $1 - z^k$, observe that the right side becomes a polynomial of degree at most $k - 1$ that is invariant under each of the k substitutions

$$z \mapsto \omega^j z \quad (j = 0, \ldots, k-1)$$

and is therefore constant, and then evaluate the constant by setting z equal to zero.

The above identity implies that if $w(A) \leq 1$ then

$$(I - z^k A^k)^{-1} = \frac{1}{k}\sum_{j=0}^{k-1}(I - \omega^j z A)^{-1}$$

whenever $|z| < 1$. Since $w(\omega^j A) \leq 1$, by (4.72) each summand on the right side has positive semidefinite real part. Hence the left side has positive semidefinite real part. Applying (4.72) again gives $w(A^k) \leq 1$. This completes the proof.   □

The above proof is, of course, very ingenious, but we can understand this result in another way. Next we will establish a theorem characterizing the unit ball of M_n with the numerical radius as the norm, so that the power inequality becomes completely obvious.

Let $\mathcal{M}$ be a subspace of M_m which contains the identity matrix and is self-adjoint, that is, $A \in \mathcal{M}$ implies $A^* \in \mathcal{M}$. A linear map $\Phi : \mathcal{M} \to M_n$ is said to be *completely positive* if for any block matrix $[X_{ij}]$ of any order with $X_{ij} \in \mathcal{M}$,

$$[X_{ij}] \geq 0 \quad \text{implies} \quad [\Phi(X_{ij})] \geq 0.$$

The following lemma is a special case of Arveson's extension theorem on C*-algebras [30, p.287].

Lemma 4.42 *Let $\mathcal{M}$ be a self-adjoint subspace of M_m containing the identity matrix. If $\Phi : \mathcal{M} \to M_n$ is a completely positive linear map, then Φ can be extended to a completely positive linear map from the whole space M_m into M_n.*

Theorem 4.43 (T. Ando) *Let $A \in M_n$. Then $w(A) \leq 1$ if and only if there are contractions $W, Z \in M_n$ with Z Hermitian such that*

$$A = (I + Z)^{1/2}W(I - Z)^{1/2}. \tag{4.73}$$

Proof. Suppose A has the expression (4.73). Then for any $u \in \mathbb{C}^n$, by the Cauchy-Schwarz inequality

$$|\langle Au, u \rangle| \leq \frac{1}{2}\{\|(I + Z)^{1/2}u\|^2 + \|(I - Z)^{1/2}u\|^2\}$$
$$= \|u\|^2.$$

Conversely suppose $w(A) \leq 1$. By Lemma 1.21 the existence of contractions W, Z with Z Hermitian satisfying the condition (4.73) is equivalent to that of a Hermitian contraction Z such that

$$\begin{bmatrix} I + Z & A \\ A^* & I - Z \end{bmatrix} \geq 0.$$

Let $\mathcal{M}$ be the complex subspace of M_2 spanned by

$$\begin{bmatrix} 1 & 0 \\ 0 & 1 \end{bmatrix}, \quad \begin{bmatrix} 0 & 1 \\ 0 & 0 \end{bmatrix}, \quad \begin{bmatrix} 0 & 0 \\ 1 & 0 \end{bmatrix}.$$

Then $\mathcal{M}$ is a self-adjoint subspace of M_2, containing the identity matrix. Its generic element is of the form

$$\begin{bmatrix} x & y \\ z & x \end{bmatrix} \quad (x, y, z \in \mathbb{C}).$$

Define a linear map Φ from $\mathcal{M}$ to M_n by

$$\Phi : \begin{bmatrix} x & y \\ z & x \end{bmatrix} \mapsto xI + \frac{1}{2}(yA + zA^*).$$

Let us show that Φ is completely positive. Let N be an arbitrary positive integer. Suppose that a matrix in $M_N \otimes \mathcal{M}$ is positive semidefinite:

$$\left(\begin{bmatrix} x_{ij} & y_{ij} \\ z_{ij} & x_{ij} \end{bmatrix} \right)_{i,j=1}^{N} \geq 0.$$

We have to show that

$$\left[x_{ij}I + \frac{1}{2}(y_{ij}A + z_{ij}A^*) \right]_{i,j=1}^{N} \geq 0.$$

In other words, we have to prove that for $X, Y \in M_N$,

$$\begin{bmatrix} X & Y \\ Y^* & X \end{bmatrix} \geq 0 \text{ implies } X \otimes I_n + \frac{1}{2}(Y \otimes A + Y^* \otimes A^*) \geq 0.$$

We may assume $X > 0$. Applying the congruence transforms with $\mathrm{diag}(X^{-1/2}, X^{-1/2})$ and $X^{-1/2} \otimes I_n$ to the above two inequalities respectively, we see that the problem is reduced to showing

$$\begin{bmatrix} I & Y \\ Y^* & I \end{bmatrix} \geq 0 \quad \text{implies} \quad I_N \otimes I_n + \frac{1}{2}(Y \otimes A + Y^* \otimes A^*) \geq 0,$$

which is equivalent to the statement that for $A \in M_n$, $Y \in M_N$,

$$w(A) \leq 1 \quad \text{and} \quad \|Y\|_\infty \leq 1 \quad \text{imply} \quad w(Y \otimes A) \leq 1.$$

Since every contraction is a convex combination of unitary matrices (in fact, by the singular value decomposition it is easy to see that every contraction Y can be written as $Y = (U_1 + U_2)/2$ with U_1, U_2 unitary), it suffices to show that if $w(A) \leq 1$ then $w(U \otimes A) \leq 1$ for unitary $U \in M_N$. To see this, consider the spectral decomposition

$$U = V \mathrm{diag}(\lambda_1, \ldots, \lambda_N) V^*$$

with V unitary and $|\lambda_j| = 1$, $j = 1, \ldots, N$. Then since $w(\cdot)$ is weakly unitarily invariant we have

$$\begin{aligned}
w(U \otimes A) &= w[(V \otimes I_n)(\mathrm{diag}(\lambda_1, \ldots, \lambda_N) \otimes A)(V \otimes I_n)^*] \\
&= w(\mathrm{diag}(\lambda_1, \ldots, \lambda_N) \otimes A) \\
&= \max_{1 \leq j \leq N} w(\lambda_j A) = w(A) \leq 1.
\end{aligned}$$

Thus we have proved that Φ is completely positive.

Now by Lemma 4.42, Φ can be extended to a completely positive map from M_2 to M_n, which we denote by the same notation Φ. Hence

$$\begin{bmatrix} \Phi\left(\begin{bmatrix} 1 & 0 \\ 0 & 0 \end{bmatrix}\right) & \Phi\left(\begin{bmatrix} 0 & 1 \\ 0 & 0 \end{bmatrix}\right) \\ \Phi\left(\begin{bmatrix} 0 & 0 \\ 1 & 0 \end{bmatrix}\right) & \Phi\left(\begin{bmatrix} 0 & 0 \\ 0 & 1 \end{bmatrix}\right) \end{bmatrix} \geq 0. \tag{4.74}$$

Since

$$\Phi\left(\begin{bmatrix} 1 & 0 \\ 0 & 0 \end{bmatrix}\right) + \Phi\left(\begin{bmatrix} 0 & 0 \\ 0 & 1 \end{bmatrix}\right) = \Phi\left(\begin{bmatrix} 1 & 0 \\ 0 & 1 \end{bmatrix}\right) = I_n,$$

there is a Hermitian matrix Z such that

$$\Phi\left(\begin{bmatrix} 1 & 0 \\ 0 & 0 \end{bmatrix}\right) = \frac{1}{2}(I + Z) \quad \text{and} \quad \Phi\left(\begin{bmatrix} 0 & 0 \\ 0 & 1 \end{bmatrix}\right) = \frac{1}{2}(I - Z).$$

On the other hand, by definition,

$$\Phi\left(\begin{bmatrix} 0 & 1 \\ 0 & 0 \end{bmatrix}\right) = \frac{1}{2}A, \quad \Phi\left(\begin{bmatrix} 0 & 0 \\ 1 & 0 \end{bmatrix}\right) = \frac{1}{2}A^*.$$

Thus (4.74) has the form

$$\begin{bmatrix} I + Z & A \\ A^* & I - Z \end{bmatrix} \geq 0.$$

This completes the proof.   □

Second Proof of Theorem 4.41. Suppose $w(A) \leq 1$. By Theorem 4.43 A has the representation

$$A = (I + Z)^{1/2} W (I - Z)^{1/2}$$

where W, Z are contractions and Z is Hermitian. Then

$$A^k = (I + Z)^{1/2} C (I - Z)^{1/2}$$

where $C = [W(I - Z^2)^{1/2}]^{k-1} W$. Obviously C is a contraction. So A^k is of the same form as A in (4.73). Applying Theorem 4.43 again in another direction yields $w(A^k) \leq 1$.   □

Notes and References. Theorem 4.41 is a conjecture of P. R. Halmos for Hilbert space operators and it was first proved by C. A. Berger [13]. Later C. Pearcy [78] gave an elementary proof. The proof given here, which is due to P. R. Halmos [42, p.320], combines the ideas of Berger and Pearcy. For generalizations of the power inequality see [59], [76], [14] and [15].

Theorem 4.43 is proved by T. Ando [2] for Hilbert space operators using an analytic method. The interesting proof using the extension theorem given here is also due to T. Ando [8].

4.7 Norm Estimates of Banded Matrices

Consider

$$A = \begin{bmatrix} 1 & 1 \\ -1 & 1 \end{bmatrix} \quad \text{and} \quad B = \begin{bmatrix} 1 & 1 \\ 0 & 1 \end{bmatrix}.$$

It is easy to see that $\|A\|_\infty = \sqrt{2}$, $\|B\|_\infty = (1 + \sqrt{5})/2$. Thus replacing an entry of a matrix by zero can increase its operator norm. On the other hand, The Frobenius norm of a matrix is diminished if any of its entries is replaced by one with smaller absolute value. It is an interesting fact that among all unitarily invariant norms, the Frobenius norm (and its constant multiples, of course) is the only one that has this property; see [21].

In this section we study the norm variations of a matrix when some of its diagonals are replaced by zeros.

A matrix $A = (a_{ij})$ is said to have *upper bandwidth* q if $a_{ij} = 0$ whenever $j - i > q$; it has *lower bandwidth* p if $a_{ij} = 0$ whenever $i - j > p$. For example, tridiagonal matrices have both upper and lower bandwidths 1; upper triangular matrices have lower bandwidth 0.

For $A \in M_n$, let $\mathfrak{D}_0(A)$ be the diagonal part of A, i.e., the matrix obtained from A by replacing all its off-diagonal entries by zeros. For $1 \leq j \leq n - 1$, let $\mathfrak{D}_j(A)$ be the matrix obtained from A by replacing all its entries except those on the jth superdiagonal by zeros. Likewise, let $\mathfrak{D}_{-j}(A)$ be the matrix obtained by retaining only the jth subdiagonal of A.

Let $\omega = e^{2\pi i/n}$, $i = \sqrt{-1}$ and let U be the diagonal matrix with entries $1, \omega, \omega^2, \ldots, \omega^{n-1}$ down its diagonal. Then we have

$$\mathfrak{D}_0(A) = \frac{1}{n} \sum_{j=0}^{n-1} U^j A U^{*j}. \tag{4.75}$$

From this expression it follows immediately that for any weakly unitarily invariant norm

$$\|\mathfrak{D}_0(A)\| \le \|A\|. \tag{4.76}$$

Next comes the key idea. The formula (4.75) represents the main diagonal as an average over unitary conjugates of A. It turns out that this idea can be generalized by considering a continuous version, i.e., using integrals, so that every diagonal has a similar expression. For each real number θ, let U_θ be the diagonal matrix with entries $e^{ir\theta}$, $1 \le r \le n$, down its diagonal. Then the (r, s) entry of the matrix $U_\theta A U_\theta^*$ is $e^{i(r-s)\theta} a_{rs}$. Hence we have

$$\mathfrak{D}_k(A) = \frac{1}{2\pi} \int_{-\pi}^{\pi} e^{ik\theta} U_\theta A U_\theta^* \, d\theta. \tag{4.77}$$

Note that when $k = 0$, this gives another representation of $\mathfrak{D}_0(A)$. From (4.77) we get

$$\|\mathfrak{D}_k(A)\| \le \|A\|, \quad 1 - n \le k \le n - 1 \tag{4.78}$$

for any weakly unitarily invariant norm. In the sequel all norms are weakly unitarily invariant unless otherwise stated.

We can also use (4.77) to write

$$\mathfrak{D}_k(A) + \mathfrak{D}_{-k}(A) = \frac{1}{2\pi} \int_{-\pi}^{\pi} (2\cos k\theta) U_\theta A U_\theta^* \, d\theta.$$

Hence

$$\|\mathfrak{D}_k(A) + \mathfrak{D}_{-k}(A)\| \le \frac{1}{2\pi} \int_{-\pi}^{\pi} |2\cos k\theta| d\theta \|A\|.$$

It is easy to evaluate this integral. We get

$$\|\mathfrak{D}_k(A) + \mathfrak{D}_{-k}(A)\| \le \frac{4}{\pi} \|A\|. \tag{4.79}$$

Let $\mathcal{T}_3(A) = \mathfrak{D}_{-1}(A) + \mathfrak{D}_0(A) + \mathfrak{D}_1(A)$ be the tridiagonal part of A. Using the same argument we see that

$$\|\mathcal{T}_3(A)\| \le \frac{1}{2\pi} \int_{-\pi}^{\pi} |1 + 2\cos\theta| d\theta \|A\|.$$

Once again, it is easy to evaluate the integral. We now get

$$\|\mathcal{T}_3(A)\| \le \left(\frac{1}{3} + \frac{2\sqrt{3}}{\pi}\right) \|A\|. \tag{4.80}$$

The constant factor in the inequality (4.80) is smaller than 1.436, and that in (4.79) is smaller than 1.274.

More generally, consider the *trimming* of A obtained by replacing all its diagonals outside the band $-k \le j \le k$ by zeros; i.e., consider the matrices

$$\mathcal{T}_{2k+1}(A) \equiv \sum_{j=-k}^{k} \mathcal{D}_j(A), \quad 1 \le k \le n-1.$$

In a similar way we can derive

$$\|\mathcal{T}_{2k+1}(A)\| \le L_k \|A\| \tag{4.81}$$

where the number L_k is the Lebesgue constant:

$$L_k = \frac{1}{2\pi} \int_{-\pi}^{\pi} |\sum_{j=-k}^{k} e^{ij\theta}| d\theta.$$

It is known that

$$L_k \le \log k + \log \pi + \frac{2}{\pi}\left(1 + \frac{1}{2k}\right),$$

and that

$$L_k = \frac{4}{\pi^2} \log k + O(1).$$

Let Δ be the linear map on the space of matrices of a fixed size that takes a matrix B to its upper triangular part; i.e., Δ acts by replacing all entries of a matrix below the main diagonal by zeros. Given a $k \times k$ matrix B, consider the matrix $A = \begin{bmatrix} 0 & B^* \\ B & 0 \end{bmatrix}$. Then

$$\mathcal{T}_{2k+1}(A) = \begin{bmatrix} 0 & \Delta(B)^* \\ \Delta(B) & 0 \end{bmatrix}.$$

Since the singular values of A are the singular values of B counted twice as often, from (4.81) via the Fan dominance principle we obtain

$$\|\Delta(B)\| \le L_k \|B\|$$

for every unitarily invariant norm.

Now let us show that the bounds (4.79) and (4.80) are sharp for the trace norm in an asymptotic sense and therefore, by a duality argument they are sharp for the operator norm.

Let $A = E$, the $n \times n$ matrix with all entries equal to 1. Then

$$\mathfrak{D}_1(A) + \mathfrak{D}_{-1}(A) = B$$

where B is the tridiagonal matrix with each entry on its first superdiagonal and the first subdiagonal equal to 1, and all other entries equal to 0. The eigenvalues of B are $2\cos(j\pi/n+1)$, $j = 1, \ldots, n$ [17, p.60]. Here

$$\frac{\|B\|_1}{\|A\|_1} = \frac{1}{n} \sum_{j=1}^{n} \left| 2\cos \frac{j\pi}{n+1} \right|. \qquad (4.82)$$

Let $f(\theta) = |2\cos\theta|$. The sum

$$\frac{1}{n+1} \sum_{j=1}^{n+1} \left| 2\cos \frac{j\pi}{n+1} \right|$$

is a Riemann sum for the function $\pi^{-1}f(\theta)$ over a subdivision of the interval $[0, \pi]$ into $n+1$ equal parts. As $n \to \infty$, this sum and the one in (4.82) tend to the same limit

$$\frac{1}{\pi} \int_0^\pi |2\cos\theta| d\theta = \frac{1}{2\pi} \int_{-\pi}^\pi |2\cos\theta| d\theta = \frac{4}{\pi}.$$

This shows that the inequality (4.79) can not be improved if it has to be valid for all dimensions n and for all unitarily invariant norms.

The same example shows that the inequality (4.80) is also sharp.

Notes and References. This section is taken from R. Bhatia [18]. S. Kwapien and A. Pelczynski proved in [65] that the norm of the triangular truncation operator Δ on $(M_n, \|\cdot\|_\infty)$ grows like $\log n$ and they gave some applications of this estimate.

5. Solution of the van der Waerden Conjecture

Let $A = (a_{ij})$ be an $n \times n$ matrix. The *permanent* of A is defined by

$$\text{per}\,A = \sum_{\sigma \in S_n} \prod_{i=1}^{n} a_{i\sigma(i)}$$

where S_n denotes the symmetric group of degree n, i.e., the group of all permutations of $\mathcal{N} \equiv \{1, 2, \ldots, n\}$. We will consider the columns $a_1, \ldots, a_n$ of A as vectors in $\mathbb{C}^n$ and write

$$\text{per}\,A = \text{per}(a_1, a_2, \ldots, a_n).$$

It is clear that $\text{per}\,A$ is a linear function of a_j for each j and we have the following expansion formulae according to columns or rows:

$$\text{per}\,A = \sum_{i=1}^{n} a_{ij}\,\text{per}\,A(i|j) = \sum_{j=1}^{n} a_{ij}\,\text{per}\,A(i|j) \tag{5.1}$$

for all i and j, where $A(i|j)$ denotes the matrix obtained from A by deleting the ith row and jth column. The permanent function is *permutation-invariant* in the sense that

$$\text{per}\,A = \text{per}(PAQ)$$

for all permutation matrices P and Q.

Recall that a square entrywise nonnegative matrix is called *doubly stochastic* if all its row and column sums are one. Let J_n be the $n \times n$ matrix whose all entries are $1/n$.

The purpose of this chapter is to prove the following theorem.

Theorem 5.1 *If A is an $n \times n$ doubly stochastic matrix, then*

$$\text{per}\,A \geq n!/n^n, \tag{5.2}$$

with equality if and only if $A = J_n$.

The assertion in Theorem 5.1 is known as *the van der Waerden conjecture* posed in 1926 [83]. It had been a long-standing open problem which, according to D. E. Knuth [63], had resisted the attacks of many strong mathematicians. Around 1980 this famous beautiful conjecture was proved independently by G. P. Egorychev [33] and D. I. Falikman [35]. Their proofs are different but both use a special case of the Alexandroff-Fenchel inequality on the geometry of convex sets [1, 36]. Falikman's method proves the inequality (5.2) without showing that J_n is the unique minimizing matrix.

In the course of looking for a proof of the van der Waerden conjecture, M. Marcus, M. Newman, and D. London made key contributions. Let Ω_n be the set of all $n \times n$ doubly stochastic matrices. Call a matrix $A \in \Omega_n$ satisfying

$$\text{per} A = \min_{X \in \Omega_n} \text{per} X$$

a *minimizing matrix*. Marcus and Newman hoped to prove the van der Waerden conjecture by establishing properties of minimizing matrices and then showing that only the matrix J_n satisfies these properties. This attacking strategy turns out to be correct.

Now we give a self-contained proof of Theorem 5.1 following Egorychev's argument.

Lemma 5.2 (Frobenius-König) *Let A be an $n \times n$ nonnegative matrix. Then* $\text{per} A = 0$ *if and only if A contains an $s \times t$ zero submatrix such that $s + t = n + 1$.*

Proof. For nonempty $\alpha, \beta \subseteq \mathcal{N}$ we denote by $A[\alpha|\beta]$ the submatrix of A with rows indexed by α and columns by β, and denote $\alpha^c = \mathcal{N} \backslash \alpha$. Obviously for any $n \times n$ matrix A, if $A[\alpha|\alpha^c] = 0$, then

$$\text{per} A = \text{per} A[\alpha|\alpha] \, \text{per} A[\alpha^c|\alpha^c]. \tag{5.3}$$

If A is nonnegative then for any $\alpha, \beta \subset \mathcal{N}$ with $|\alpha| + |\beta| = n$ ($|\alpha|$ means the cardinality of α),

$$\text{per} A \geq \text{per} A[\alpha|\beta^c] \, \text{per} A[\alpha^c|\beta]. \tag{5.4}$$

Now suppose $A[\alpha|\beta] = 0$ with $|\alpha| + |\beta| = n + 1$. Since the permanent is permutation-invariant, we may assume $\alpha = \{1, \ldots, k\}$, $\beta = \{k, \ldots, n\}$ for some k. Then $A[\alpha|\alpha^c] = 0$ and since the last column of $A[\alpha|\alpha]$ is 0, $\text{per} A[\alpha|\alpha] = 0$. By (5.3) we get $\text{per} A = 0$.

Conversely suppose $\text{per} A = 0$. We use induction on n. The cases $n = 1, 2$ are trivially true. Assume that the assertion is true for all orders less than n. We may assume the (n, n) entry of A, $a_{nn} > 0$. Then (5.4) implies $\text{per} A[\mathcal{N} \backslash \{n\} | \mathcal{N} \backslash \{n\}] = 0$. By the induction assumption, there exist $\alpha_1, \beta_1 \subset \{1, \ldots, n-1\}$ with $|\alpha_1| + |\beta_1| = n$ such that $A[\alpha_1|\beta_1] = 0$. Since $\text{per} A = 0$, (5.4) implies that one of $\text{per} A[\alpha_1|\beta_1^c]$ and $\text{per} A[\alpha_1^c|\beta_1]$ vanishes. We may assume $\text{per} A[\alpha_1|\beta_1^c] = 0$ (the other case is similar). Using the induction assumption again, there exist $\alpha_2 \subseteq \alpha_1$, $\beta_2 \subseteq \beta_1^c$ with $|\alpha_2| + |\beta_2| = |\alpha_1| + 1$ such that $A[\alpha_2|\beta_2] = 0$. Now

$$|\alpha_2| + |\beta_1 \cup \beta_2| = |\alpha_2| + |\beta_1| + |\beta_2| = n + 1$$

and

$$A[\alpha_2|\beta_1 \cup \beta_2] = 0.$$

This completes the proof. □

Lemma 5.2 is equivalent to the following: A necessary and sufficient condition for every generalized diagonal of an $n \times n$ arbitrary matrix A, i.e., $(a_{1\sigma(1)}, \ldots, a_{n\sigma(n)})$ for some $\sigma \in S_n$, to contain a zero entry is that A have an $s \times t$ zero submatrix such that $s + t = n + 1$.

Corollary 5.3 *The permanent of a doubly stochastic matrix is positive.*

Proof. Let A be an $n \times n$ doubly stochastic matrix. If $\mathrm{per}\,A = 0$, then, by Lemma 5.2, there exist permutation matrices P and Q such that

$$PAQ = \begin{bmatrix} B & C \\ 0 & D \end{bmatrix}$$

where the zero submatrix in the bottom left corner is $s \times t$ with $s + t = n + 1$. Denote by $\gamma(X)$ the sum of all the entries of a matrix X. Since $A \in \Omega_n$ implies $PAQ \in \Omega_n$,

$$n = \gamma(PAQ) \geq \gamma(B) + \gamma(D).$$

Now all the nonzero entries in the first t columns are contained in B, and thus $\gamma(B) = t$. Similarly $\gamma(D) = s$. Hence

$$n \geq \gamma(B) + \gamma(D) = s + t$$

which is impossible, since $s + t = n + 1$. □

An $n \times n$ matrix is said to be *partly decomposable* if it contains a $k \times (n - k)$ zero submatrix for some $1 \leq k \leq n - 1$. In other words, there exist permutation matrices P and Q such that

$$PAQ = \begin{bmatrix} B & C \\ 0 & D \end{bmatrix}$$

where B and D are square. If A is not partly decomposable, it is called *fully indecomposable*.

Corollary 5.4 *A nonnegative n-square ($n \geq 2$) matrix A is fully indecomposable if and only if*

$$\mathrm{per}\,A(i|j) > 0 \quad \text{for all } i \text{ and } j.$$

Proof. By Lemma 5.2, $\mathrm{per}\,A(i|j) = 0$ for some i, j if and only if the submatrix $A(i|j)$ and hence A contains an $s \times t$ zero submatrix with $s + t = n$. In other words, $\mathrm{per}\,A(i|j) = 0$ for some pair i, j if and only if A is partly decomposable. $\square$

Lemma 5.5 *If A is a partly decomposable doubly stochastic matrix, then there exist permutation matrices P, Q such that PAQ is a direct sum of doubly stochastic matrices.*

Proof. Since A is partly decomposable, there exist permutation matrices P and Q such that

$$PAQ = \begin{bmatrix} B & C \\ 0 & D \end{bmatrix}$$

where B and D are square. Using the same argument as in the proof of Corollary 5.3 it is easy to show $C = 0$. $\square$

Denote by $e \in \mathbb{R}^n$ the vector with all components equal to 1. Then the condition that all the row and column sums of an n-square matrix A be 1 can be expressed as $Ae = e$ and $A^T e = e$. From this it is obvious that if A is doubly stochastic then so are $A^T A$ and AA^T.

Lemma 5.6 *If A is a fully indecomposable doubly stochastic matrix, then so are $A^T A$ and AA^T.*

Proof. We have already remarked that both $A^T A$ and AA^T are doubly stochastic. Suppose $A^T A$ is partly decomposable and A is $n \times n$, i.e., there exist nonempty $\alpha, \beta \subset \mathcal{N}$ such that

$$|\alpha| + |\beta| = n \quad and \quad A^T A[\alpha|\beta] = 0.$$

The latter condition is equivalent to

$$(A[\mathcal{N}|\alpha])^T A[\mathcal{N}|\beta] = 0,$$

which implies that the sum of the numbers of zero rows of $A[\mathcal{N}|\alpha]$ and of $A[\mathcal{N}|\beta]$ is at least n.

Let t be the number of zero rows of $A[\mathcal{N}|\alpha]$. If $t \geq n - |\alpha|$, then $A[\mathcal{N}|\alpha]$ and hence A has an $|\alpha| \times (n - |\alpha|)$ zero submatrix, which contradicts the assumption that A is fully indecomposable. But if $t < n - |\alpha| (= |\beta|)$, the number s of zero rows of $A[\mathcal{N}|\beta]$ satisfies $s \geq n - t > n - |\beta|$, so that A is partly decomposable, which again contradicts the assumption.

In the same way we can show that AA^T is fully indecomposable. $\square$

Lemma 5.7 *If A is a fully indecomposable doubly stochastic matrix and $Ax = x$ for some real vector x, then x is a constant multiple of e.*

Proof. Suppose, to the contrary, that x has some components unequal. Since A is doubly stochastic, $Ax = x$ implies $A(x+\xi e) = x+\xi e$ for any real number ξ. Thus we may assume that all components of x are nonnegative and some are zero. Let $x = (x_i)$, $A = (a_{ij})$ and $\alpha = \{i : x_i = 0\}$, $\beta = \{j : x_j > 0\}$. Then

$$\sum_j a_{ij} x_j = x_i \quad \text{implies} \quad a_{ij} = 0$$

whenever $i \in \alpha$ and $j \in \beta$. Hence A is partly decomposable, contradicting the assumption. $\qquad\square$

Lemma 5.8 (M. Marcus and M. Newman [70]) *A minimizing matrix is fully indecomposable.*

Proof. Let $A \in \Omega_n$ be a minimizing matrix. Suppose that A is partly decomposable. Then by Lemma 5.5 there exist permutation matrices P and Q such that $PAQ = B \oplus C$ where $B = (b_{ij}) \in \Omega_k$ and $C = (c_{ij}) \in \Omega_{n-k}$. We will show that there exists a doubly stochastic matrix whose permanent is less than per$A$.

Denote $\tilde{A} \equiv PAQ$. Since per$A = $ per$\tilde{A} > 0$ by Corollary 5.3, using the expansion formula (5.1) and the permutation-invariance of the permanent function we can assume without loss of generality that

$$b_{kk}\mathrm{per}\tilde{A}(k|k) > 0 \quad \text{and} \quad c_{11}\mathrm{per}\tilde{A}(k+1|k+1) > 0.$$

Let ϵ be any positive number smaller than $\min(b_{kk}, c_{11})$ and define

$$G(\epsilon) \equiv \tilde{A} - \epsilon(E_{kk} + E_{k+1,k+1}) + \epsilon(E_{k,k+1} + E_{k+1,k})$$

where E_{ij} is the matrix whose only nonzero entry is 1 in (i,j). Then $G(\epsilon) \in \Omega_n$ and

$$\begin{aligned}
\mathrm{per}G(\epsilon) &= \mathrm{per}\tilde{A} - \epsilon\mathrm{per}\tilde{A}(k|k) + \epsilon\mathrm{per}\tilde{A}(k|k+1) \\
&\quad -\epsilon\mathrm{per}\tilde{A}(k+1|k+1) + \epsilon\mathrm{per}\tilde{A}(k+1|k) + O(\epsilon^2) \\
&= \mathrm{per}A - \epsilon[\mathrm{per}\tilde{A}(k|k) + \mathrm{per}\tilde{A}(k+1|k+1)] + O(\epsilon^2),
\end{aligned}$$

since per$\tilde{A}(k|k+1) = $ per$\tilde{A}(k+1|k) = 0$ by Lemma 5.2. Also,

$$\mathrm{per}\tilde{A}(k|k) + \mathrm{per}\tilde{A}(k+1|k+1) > 0$$

and therefore for sufficiently small positive ϵ

$$\mathrm{per}G(\epsilon) < \mathrm{per}A,$$

contradicting the assumption that A is a minimizing matrix. $\qquad\square$

Lemma 5.9 (M. Marcus and M. Newman [70]) *If $A = (a_{ij})$ is a minimizing matrix and $a_{hk} > 0$, then*

$$\operatorname{per} A(h|k) = \operatorname{per} A.$$

Proof. Let $C(A)$ be the face of the convex set Ω_n of least dimension containing A in its interior, i.e.,

$$C(A) \equiv \{X = (x_{ij}) \in \Omega_n : x_{ij} = 0 \text{ if } (i,j) \in \Delta\}$$

where $\Delta \equiv \{(i,j) : a_{ij} = 0\}$. Then $C(A)$ is defined by the following conditions:

$$x_{ij} \geq 0, \ i,j = 1, \ldots, n; \quad x_{ij} = 0, \ (i,j) \in \Delta;$$

$$\sum_{j=1}^{n} x_{ij} = 1, \ i = 1, \ldots, n; \quad \sum_{i=1}^{n} x_{ij} = 1, \ j = 1, \ldots, n.$$

According to the Lagrange multiplier method, there exist $\lambda = (\lambda_1, \ldots, \lambda_n)^T$, $\mu = (\mu_1, \ldots, \mu_n)^T \in \mathbb{R}^n$ such that the real function f defined on $C(A)$ by

$$f(X) = \operatorname{per} X - \sum_{i=1}^{n} \lambda_i \left(\sum_{k=1}^{n} x_{ik} - 1 \right) - \sum_{j=1}^{n} \mu_j \left(\sum_{k=1}^{n} x_{kj} - 1 \right)$$

attains its local minimum at A. Therefore for $(i,j) \notin \Delta$

$$0 = \left. \frac{\partial f(X)}{\partial x_{ij}} \right|_{X=A} = \operatorname{per} A(i|j) - \lambda_i - \mu_j. \tag{5.5}$$

In view of (5.1), (5.5) implies

$$\operatorname{per} A = \lambda_i + \sum_{j=1}^{n} a_{ij} \mu_j \quad \text{for all } i,$$

$$\operatorname{per} A = \sum_{i=1}^{n} \lambda_i a_{ij} + \mu_j \quad \text{for all } j$$

which can be written equivalently as

$$(\operatorname{per} A)e = \lambda + A\mu,$$

$$(\operatorname{per} A)e = A^T \lambda + \mu.$$

These two relations, together with $A^T e = e$ and $Ae = e$, yield

$$A^T A \mu = \mu, \quad AA^T \lambda = \lambda. \tag{5.6}$$

By Lemma 5.8, A is fully indecomposable. Therefore both $A^T A$ and AA^T are fully indecomposable and doubly stochastic by Lemma 5.6. Then applying Lemma 5.7 to (5.6) we deduce that both λ and μ are constant multiples of e, say, $\lambda = ce$ and $\mu = de$. It follows from (5.5) that

$$\mathrm{per}\, A(i|j) = c + d \quad \text{for all} \ (i,j) \notin \Delta.$$

Hence

$$\mathrm{per}\, A = \sum_{j=1}^{n} a_{ij}\, \mathrm{per}\, A(i|j)$$

$$= \sum_{j=1}^{n} a_{ij}(c+d)$$

$$= c + d$$

$$= \mathrm{per}\, A(i|j)$$

for all $(i,j) \notin \Delta$. $\square$

Lemma 5.10 (D. London [69]) *If A is a minimizing matrix then*

$$\mathrm{per}\, A(i|j) \geq \mathrm{per}\, A \quad \text{for all} \ i \ \text{and} \ j.$$

Proof. Let $P = (p_1, \ldots, p_n)$ be the permutation matrix corresponding to a permutation $\sigma \in S_n$, i.e., $p_j = e_{\sigma(j)}$ where $e_1, \ldots, e_n$ are the standard basis for $\mathbb{R}^n$. For $0 \leq \theta \leq 1$ define

$$\psi_P(\theta) = \mathrm{per}((1-\theta)A + \theta P).$$

Since $\mathrm{per}\, A$ is a linear function of each column of $A = (a_1, \ldots, a_n)$,

$$\psi_P(\theta) = \mathrm{per}((1-\theta)A + \theta P)$$
$$= (1-\theta)^n \mathrm{per}\, A + \theta^2 \varphi(\theta)$$
$$+ (1-\theta)^{n-1}\theta \sum_{t=1}^{n} \mathrm{per}(a_1, \ldots, a_{t-1}, p_t, a_{t+1}, \ldots, a_n)$$

where $\varphi(\theta)$ is a polynomial in θ. Thus

$$\psi_P'(0) = \sum_{t=1}^{n} \mathrm{per}\, A(\sigma(t)|t) - n\, \mathrm{per}\, A. \tag{5.7}$$

Since $\psi_P(0) = \mathrm{per}\, A$ is the minimum value and $(1-\theta)A + \theta P \in \Omega_n$, $\psi_P'(0) \geq 0$. From (5.7) we get

$$\sum_{t=1}^{n} \mathrm{per}\, A(\sigma(t)|t) \geq n\, \mathrm{per}\, A \tag{5.8}$$

for all $\sigma \in S_n$.

Lemma 5.8 asserts that A is fully indecomposable. Then it follows from Corollary 5.4 that for every pair i, j, $\mathrm{per}\, A(i|j) > 0$. Hence there is a permutation σ such that $i = \sigma(j)$ and $a_{\sigma(t),t} > 0$ for $1 \leq t \leq n$, $t \neq j$. By Lemma 5.9 we have

$$\mathrm{per}A(\sigma(t)|t) = \mathrm{per}A, \quad 1 \le t \le n, \, t \ne j. \tag{5.9}$$

Combining (5.8) and (5.9) yields $\mathrm{per}A(i|j) \ge \mathrm{per}A$ for all i, j. $\qquad\square$

Next we derive another key ingredient of the proof of van der Waerden's conjecture: the Alexandroff-Fenchel inequality. A real $n \times n$ symmetric matrix Q defines a symmetric inner product $\langle x, y \rangle_Q \equiv x^T Q y$ on $\mathbb{R}^n$. If Q has one positive eigenvalue and $n-1$ negative eigenvalues, then the space $\mathbb{R}^n$ equipped with the inner product defined by Q is called a *Lorentz space*. We call a vector $x \in \mathbb{R}^n$ *positive* if $\langle x, x \rangle_Q > 0$.

Lemma 5.11 *Let a be a positive vector in a Lorentz space with a symmetric inner product $\langle \cdot, \cdot \rangle_Q$. Then for any $b \in \mathbb{R}^n$*

$$\langle a, b \rangle_Q^2 \ge \langle a, a \rangle_Q \langle b, b \rangle_Q$$

and equality holds if and only if $b = \lambda a$ for some constant λ.

Proof. Consider the quadratic polynomial

$$f(\lambda) \equiv \langle \lambda a + b, \lambda a + b \rangle_Q = \langle a, a \rangle_Q \lambda^2 + 2 \langle a, b \rangle_Q \lambda + \langle b, b \rangle_Q.$$

We have assumed that the inner product is defined by Q, i.e.,

$$\langle x, y \rangle_Q = x^T Q y.$$

By the minimax principle for eigenvalues of Hermitian matrices, for any subspace $S \subset \mathbb{R}^n$ with $\dim S = 2$,

$$0 > \lambda_2(Q) \ge \min_{x \in S, \, \|x\|=1} \langle x, x \rangle_Q. \tag{5.10}$$

Assume $b \ne \lambda a$ for all $\lambda \in \mathbb{R}$. Then a and b spans a 2-dimensional subspace of $\mathbb{R}^n$. It follows from (5.10) that there are $\alpha, \beta \in \mathbb{R}$ such that

$$\langle \alpha a + \beta b, \alpha a + \beta b \rangle_Q < 0.$$

But since $\langle a, a \rangle_Q > 0$, $\beta \ne 0$. Thus $f(\alpha/\beta) < 0$. Therefore $f(\lambda)$ has a positive discriminant, i.e., $\langle a, b \rangle_Q^2 \ge \langle a, a \rangle_Q \langle b, b \rangle_Q$. $\qquad\square$

Given vectors $a_1, a_2, \ldots, a_{n-2}$ in $\mathbb{R}^n$ with positive components, we define an inner product on $\mathbb{R}^n$ by

$$\langle x, y \rangle_Q \equiv \mathrm{per}(a_1, a_2, \ldots, a_{n-2}, x, y), \tag{5.11}$$

that is, $\langle x, y \rangle_Q = x^T Q y$ where $Q = (q_{ij})$ is given by

$$q_{ij} = \mathrm{per}(a_1, a_2, \ldots, a_{n-2}, e_i, e_j). \tag{5.12}$$

Lemma 5.12 $\mathbb{R}^n$ *with the inner product (5.11) is a Lorentz space.*

Proof. We need to show that the symmetric matrix Q defined in (5.12) has one positive eigenvalue and $n-1$ negative eigenvalue. Use induction on n. For $n = 2$, $Q = \begin{bmatrix} 0 & 1 \\ 1 & 0 \end{bmatrix}$ and the assertion is true. Now assume that the assertion is true for $\mathbb{R}^{n-1}$ ($n \geq 3$) and consider the case n.

We first show that 0 is not an eigenvalue of Q. Suppose $Qc = 0$ for some $c \in \mathbb{R}^n$. Then $e_i^T Qc = 0$, which means

$$\mathrm{per}(a_1, \ldots, a_{n-2}, c, e_i) = 0, \quad 1 \leq i \leq n. \tag{5.13}$$

For a fixed i, by the induction hypothesis,

$$\mathrm{per}[(a_1, \ldots, a_{n-3}, x, y, e_i)(i|n)]$$

defines an inner product with which $\mathbb{R}^{n-1}$ is a Lorentz space. Applying Lemma 5.11 with the assumption that $a_1, \ldots, a_{n-2}$ have positive components and (5.13) we obtain

$$\mathrm{per}(a_1, \ldots, a_{n-3}, c, c, e_i) \leq 0 \tag{5.14}$$

for $i = 1, \ldots, n$. On the other hand, from

$$0 = c^T Qc = \langle c, c \rangle_Q = \sum_{i=1}^{n} a_{n-2}(i)\mathrm{per}(a_1, \ldots, a_{n-3}, c, c, e_i),$$

each $a_{n-2}(i) > 0$ and (5.14) we have

$$\mathrm{per}(a_1, \ldots, a_{n-3}, c, c, e_i) = 0 \quad 1 \leq i \leq n. \tag{5.15}$$

Therefore the equality case in (5.14) holds. By Lemma 5.11, for each i there is a number λ_i such that $c = \lambda_i a_{n-2}$ except the ith component. Since $n \geq 3$, this is possible only when all λ_i, $i = 1, \ldots, n$ are equal. So, in fact, $c = \lambda a_{n-2}$. In view of (5.15) we must have $\lambda = 0$, that is, $c = 0$. This proves that Q is nonsingular, or equivalently, 0 is not an eigenvalue.

Next consider the matrix Q_θ whose (i, j) entry is given by

$$\mathrm{per}(\theta a_1 + (1 - \theta)e, \ldots, \theta a_{n-2} + (1 - \theta)e, e_i, e_j)$$

where $e = (1, 1, \ldots, 1)^T$. Then for each $\theta \in [0, 1]$, by what we have just proved, 0 is not an eigenvalue of Q_θ. Since eigenvalues are continuous functions of matrix entries, $Q = Q_1$ and Q_0 have the same number of positive eigenvalues. It is easy to see that Q_0 has exactly one positive eigenvalue. This completes the proof. □

Lemma 5.13 (Alexandroff-Fenchel) *Let $a_1, a_2, \ldots, a_{n-1}$ be vectors in $\mathbb{R}^n$ with positive components. Then for any $b \in \mathbb{R}^n$*

$$(\text{per}(a_1, \ldots, a_{n-1}, b))^2$$
$$\geq \text{per}(a_1, \ldots, a_{n-1}, a_{n-1})\text{per}(a_1, \ldots, a_{n-2}, b, b), \qquad (5.16)$$

with equality if and only if $b = \lambda a_{n-1}$ for some constant λ. The inequality (5.16) itself also holds when $a_1, a_2, \ldots, a_{n-1}$ have nonnegative components.

Proof. This is a consequence of Lemmas 5.12 and 5.11. $\qquad\square$

Lemma 5.14 *If A is a minimizing matrix, then*

$$\text{per}\,A(i|j) = \text{per}\,A \quad \text{for all } i \text{ and } j.$$

Proof. Let $A = (a_{ij}) = (a_1, \ldots, a_n)$. Suppose the assertion is false. Then by Lemma 5.10, there is a pair r, s such that $\text{per}\,A(r|s) > \text{per}\,A \ (> 0)$. Since A is fully indecomposable by Lemma 5.8, every row of A has at least two positive entries. Hence there is a $t \neq s$ such that $a_{rt} > 0$.

Now apply Lemma 5.13 to get

$$(\text{per}\,A)^2 = [\text{per}(a_1, \ldots, a_s, \ldots, a_t, \ldots, a_n)]^2$$
$$\geq \text{per}(a_1, \ldots, a_s, \ldots, a_s, \ldots, a_n)\text{per}(a_1, \ldots, a_t, \ldots, a_t, \ldots, a_n)$$
$$= \left(\sum_{k=1}^{n} a_{ks}\text{per}\,A(k|t)\right)\left(\sum_{k=1}^{n} a_{kt}\text{per}\,A(k|s)\right). \qquad (5.17)$$

Every subpermanent above is at least $\text{per}\,A$ by Lemma 5.10 and $\text{per}\,A(r|s) > \text{per}\,A$. Since $\text{per}\,A(r|s)$ is multiplied by a_{rt}, which is positive, the last product in (5.17) is greater than $(\text{per}\,A)^2$, contradicting the inequality (5.17). $\qquad\square$

Lemma 5.15 *If $A = (a_1, \ldots, a_n)$ is a minimizing matrix and A' is the matrix obtained from A by replacing both a_i and a_j by $(a_i + a_j)/2$ for any pair i, j, then A' is again a minimizing matrix.*

Proof. Obviously A' is doubly stochastic. By (5.1) and Lemma 5.14 we have

$$\text{per}\,A' = \frac{1}{2}\text{per}\,A + \frac{1}{4}\text{per}(a_1, \ldots, a_i, \ldots, a_i, \ldots, a_n)$$
$$+ \frac{1}{4}\text{per}(a_1, \ldots, a_j, \ldots, a_j, \ldots, a_n)$$
$$= \frac{1}{2}\text{per}\,A + \frac{1}{4}\sum_{k=1}^{n} a_{ki}\text{per}\,A(k|j) + \frac{1}{4}\sum_{k=1}^{n} a_{kj}\text{per}\,A(k|i)$$
$$= \text{per}\,A.$$

$\qquad\square$

Proof of Theorem 5.1. Let $A = (a_1, \ldots, a_n)$ be a minimizing matrix. We consider an arbitrary column of A, say, a_n. Since A is fully indecomposable by Lemma 5.8, every row of A has at least two positive entries. Hence a finite number of applications of Lemma 5.15 to the columns $a_1, \ldots, a_{n-1}$ yields a minimizing matrix A' which also has a_n as its last column and whose other columns $a_1', \ldots, a_{n-1}'$ all have positive components.

Now apply Lemma 5.13 to $\mathrm{per}(a_1', \ldots, a_{n-1}', a_n)$. By expanding the permanents on both sides of the inequality using Lemma 5.14, we see that equality holds. It follows that a_n is a multiple of a_{n-1}' and similarly we find that a_n is a multiple of a_i' for all $1 \leq i \leq n-1$. But since the sum of the components of each of $a_1', \ldots, a_{n-1}', a_n$ is 1, we obtain $a_n = a_i'$, $i = 1, \ldots, n-1$. Thus $e = \sum_{i=1}^{n-1} a_i' + a_n = na_n$ and hence $a_n = n^{-1}e$.

In the same way we can show that each column of A is $n^{-1}e$, that is, $A = J_n$. This completes the proof. $\qquad\square$

Notes and References. The presentation of the proofs given here follows the expositions by H. Minc [75], J. H. van Lint [67] and T. Ando [5]. For the history of the van der Waerden conjecture see [75] and [68].

References

1. A. D. Alexandroff, *Zur Theorie der gemischten Volumina von Konvexen Körpern IV*, Mat. Sbornik **3**(45) (1938) 227-251.
2. T. Ando, *Structure of operators with numerical radius one*, Acta Sci. Math (Szeged), **34**(1973) 11-15.
3. T. Ando, *Concavity of certain maps on positive definite matrices and applications to Hadamard products*, Linear Algebra Appl., **26**(1979) 203-241.
4. T. Ando, *Comparison of norms $|||f(A) - f(B)|||$ and $|||f(|A - B|)|||$*, Math. Z., **197**(1988) 403-409.
5. T. Ando, *Majorizations, doubly stochastic matrices, and comparison of eigenvalues*, Linear Algebra Appl., **118**(1989) 163-248.
6. T. Ando, *Majorization relations for Hadamard products*, Linear Algebra Appl., **223/224**(1995) 57-64.
7. T. Ando, *Matrix Young inequalities*, Operator Theory: Advances and Applications, **75**(1995) 33-38.
8. T. Ando, *Operator-Theoretic Methods for Matrix Inequalities*, Hokusei Gakuen Univ., 1998.
9. T. Ando and R. Bhatia, *Eigenvalue inequalities associated with the Cartesian decomposition*, Linear and Multilinear Algebra, **22**(1987) 133-147.
10. T. Ando and F. Hiai, *Hölder type inequalities for matrices*, Math. Ineq. Appl., **1**(1998) 1-30.
11. T. Ando and X. Zhan, *Norm inequalities related to operator monotone functions*, Math. Ann., **315**(1999) 771-780.
12. R. B. Bapat and V. S. Sunder, *On majorization and Schur products*, Linear Algebra Appl., **72**(1985) 107-117.
13. C. A. Berger, *Abstract 625-152*, Notices Amer. Math. Soc., **12**(1965) 590.
14. C. A. Berger and J. G. Stampfli, *Norm relations and skew dilations*, Acta Sci. Math. (Szeged), **28**(1967) 191-195.
15. C. A. Berger and J. G. Stampfli, *Mapping theorems for the numerical range*, Amer. J. Math., **89**(1967) 1047-1055.
16. R. Bhatia, *Perturbation Bounds for Matrix Eigenvalues*, Longman, 1987.
17. R. Bhatia, *Matrix Analysis*, GTM 169, Springer-Verlag, New York, 1997.
18. R. Bhatia, *Pinching, trimming, truncating, and averaging of matrices*, Amer. Math. Monthly, **107**(2000) 602-608.
19. R. Bhatia and C. Davis, *More matrix forms of the arithmetic-geometric mean inequality*, SIAM J. Matrix Anal. Appl., **14** (1993) 132-136 .
20. R. Bhatia and C. Davis, *A Cauchy-Schwarz inequality for operators with applications*, Linear Algebra Appl., **223/224**(1995) 119-129.

21. R. Bhatia, C. Davis and M. D. Choi, *Comparing a matrix to its off-diagonal part*, Operator Theory: Advances and Applications, **40**(1989) 151-164.

22. R. Bhatia and F. Kittaneh, *On the singular values of a product of operators*, SIAM J. Matrix Anal. Appl., **11**(1990) 272-277.

23. R. Bhatia and F. Kittaneh, *Norm inequalities for positive operators*, Lett. Math. Phys., **43**(1998) 225-231.

24. R. Bhatia and F. Kittaneh, *Notes on matrix arithmetic-geometric mean inequalities*, Linear Algebra Appl., **308**(2000) 203-211.

25. R. Bhatia and F. Kittaneh, *Cartesian decompositions and Schatten norms*, Linear Algebra Appl., **318**(2000) 109-116.

26. R. Bhatia and K. R. Parthasarathy, *Positive definite functions and operator inequalities*, Bull. London Math. Soc., **32**(2000) 214-228.

27. R. Bhatia and X. Zhan, *Compact operators whose real and imaginary parts are positive*, Proc. Amer. Math. Soc., **129**(2001) 2277-2281.

28. R. Bhatia and X. Zhan, *Norm inequalities for operators with positive real part*, J. Operator Theory, to appear.

29. N. N. Chan and M. K. Kwong, *Hermitian matrix inequalities and a conjecture*, Amer. Math. Monthly, **92**(1985) 533-541.

30. K. R. Davidson, *Nest Algebras*, Longman, 1988.

31. W. F. Donoghue, *Distributions and Fourier Transforms*, Academic Press, 1969.

32. W. F. Donoghue, *Monotone Matrix Functions and Analytic Continuation*, Springer-Verlag, 1974.

33. G. P. Egorychev, *The solution of van der Waerden's problem for permanents*, Adv. in Math., **42**(1981), 299-305.

34. L. Elsner and S. Friedland, *Singular values, doubly stochastic matrices, and applications*, Linear Algebra Appl., **220**(1995) 161-169.

35. D. I. Falikman, *A proof of the van der Waerden conjecture on the permanent of a doubly stochastic matrix*, Mathematiceski Zametki, **29**(1981) 931-938 (Russian).

36. W. Fenchel, *Inégalités quadratiques entre les volumes mixtes des corps convexes*, Comptes Rendus, Acad. Sci., Paris, **203** (1936) 647-650.

37. M. Fiedler, *A note on the Hadamard product of matrices*, Linear Algebra Appl., **49**(1983) 233-235.

38. T. Furuta, $A \geq B \geq 0$ *assures* $(B^r A^p B^r)^{1/q} \geq B^{(p+2r)/q}$ *for* $r \geq 0, p \geq 0, q \geq 1$ *with* $(1 + 2r)q \geq p + 2r$, Proc. Amer. Math. Soc., **101**(1987) 85-88.

39. T. Furuta, *Two operator functions with monotone property*, Proc. Amer. Math. Soc., **111**(1991) 511-516.

40. I. C. Gohberg and M. G. Krein, *Introduction to the Theory of Linear Nonselfadjoint Operators*, AMS, Providence, RI, 1969.

41. I. S. Gradshteyn and I. M. Ryzhik, *Tables of Integrals, Series, and Products*, 5th Ed., Academic Press, New York, 1994.

42. P. R. Halmos, *A Hilbert Space Problem Book*, GTM 19, 2nd Ed., Springer-Verlag, 1982.

43. F. Hansen, *An operator inequality*, Math. Ann., **246**(1980) 249-250.

44. F. Hansen and G. K. Pedersen, *Jensen's inequality for operators and Löwner's theorem*, Math. Ann, **258**(1982) 229-241.

45. H. Helson, *Harmonic Analysis*, 2nd Ed., Hindustan Book Agency, Delhi, 1995.

46. F. Hiai, *Log-majorizations and norm inequalities for exponential operators*, Banach Center Publications, Vol. **38**, pp.119-181, 1997.

47. F. Hiai and H. Kosaki, *Comparison of various means for operators*, J. Funct. Anal., **163**(1999) 300-323.

48. F. Hiai and H. Kosaki, *Means for matrices and comparison of their norms*, Indiana Univ. Math. J., **48**(1999) 899-936.

49. F. Hiai and X. Zhan, *Inequalities involving unitarily invariant norms and operator monotone functions*, Linear Algebra Appl., **341**(2002), 151-169.

50. R. A. Horn, *Norm bounds for Hadamard products and the arithmetic-geometric mean inequality for unitarily invariant norms*, Linear Algebra Appl., **223/224**(1995), 355-361.

51. R. A. Horn and C. R. Johnson, *Matrix Analysis*, Cambridge University Press, 1985.

52. R. A. Horn and C. R. Johnson, *Topics in Matrix Analysis*, Cambridge University Press, 1991.

53. R. A. Horn and R. Mathias, *Cauchy-Schwarz inequalities associated with positive semidefinite matrices*, Linear Algebra Appl., **142**(1990) 63-82.

54. R. A. Horn and R. Mathias, *An analog of the Cauchy-Schwarz inequality for Hadamard products and unitarily invariant norms*, SIAM J. Matrix Anal. Appl., **11**(1990) 481-498.

55. R. A. Horn and X. Zhan, *Inequalities for C-S seminorms and Lieb functions*, Linear Algebra Appl., **291**(1999) 103-113.

56. K. D. Ikramov, *A simple proof of the generalized Schur inequality*, Linear Algebra Appl., **199**(1994) 143-149.

57. C. R. Johnson and R. B. Bapat, *A weak multiplicative majorization conjecture for Hadamard products*, Linear Algebra Appl., **104**(1988) 246-247.

58. C. R. Johnson and L. Elsner, *The relationship between Hadamard and conventional multiplications for positive definite matrices*, Linear Algebra Appl., **92**(1987) 231-240.

59. T. Kato, *Some mapping theorems for the numerical range*, Proc. Japan Acad. **41**(1965) 652-655.

60. T. Kato, *Spectral order and a matrix limit theorem*, Linear and Multilinear Algebra, **8**(1979) 15-19.

61. F. Kittaneh, *A note on the arithmetic-geometric mean inequality for matrices*, Linear Algebra Appl., **171**(1992) 1-8.

62. F. Kittaneh, *Norm inequalities for fractional powers of positive operators*, Lett. Math. Phys., **27**(1993) 279-285.

63. D. E. Knuth, *A permanent inequality*, Amer. Math. Monthly, **88**(1981) 731-740.

64. H. Kosaki, *Arithmetic-geometric mean and related inequalities for operators*, J. Funct. Anal., **156**(1998) 429-451.

65. S. Kwapien and A. Pelczynski, *The main triangle projection in matrix spaces and its applications*, Studia Math., **34**(1970) 43-68.

66. M. K. Kwong, *Some results on matrix monotone functions*, Linear Algebra Appl., **118**(1989) 129-153.

67. J. H. van Lint, *Notes on Egoritsjev's proof of the van der Waerden conjecture*, Linear Algebra Appl., **39**(1981) 1-8.

68. J. H. van Lint, *The van der Waerden conjecture: two proofs in one year*, Math. Intelligencer, **4**(1982) 72-77.

69. D. London, *Some notes on the van der Waerden conjecture*, Linear Algebra Appl., **4**(1971) 155-160.

70. M. Marcus and M. Newman, *On the minimum of the permanent of a doubly stochastic matrix*, Duke Math. J. **26**(1959) 61-72.

71. A. W. Marshall and I. Olkin, *Norms and inequalities for condition numbers*, Pacific J. Math., **15**(1965) 241-247.

72. A. W. Marshall and I. Olkin, *Inequalities: Theory of Majorization and Its Applications*, Academic Press, 1979.

73. R. Mathias, *An arithmetic-geometric-harmonic mean inequality involving Hadamard products*, Linear Algebra Appl., **184** (1993) 71-78.

74. A. McIntosh, *Heinz inequalities and perturbation of spectral families*, Macquarie Mathematical Reports 79-0006, 1979.

75. H. Minc, *Permanents*, Addison-Wesley, Reading, MA, 1978.

76. B. Sz-Nagy and C. Foias, *On certain classes of power-bounded operators in Hilbert space*, Acta Szeged **27**(1966) 17-25.

77. T. Ogasawara, *A theorem on operator algebras*, J. Sci. Hiroshima Univ., **18**(1955) 307-309.

78. C. Pearcy, *An elementary proof of the power inequality for the numerical radius*, Mich. Math. J., **13**(1966) 289-291.

79. G. K. Pedersen, *Some operator monotone functions*, Proc. Amer. Math. Soc., **36**(1972) 309-310.

80. J. F. Queiró and A. L. Duarte, *On the Cartesian decomposition of a matrix*, Linear and Multilinear Algebra, **18**(1985) 77-85.

81. E. Stein and G. Weiss, *Introduction to Fourier Analysis on Euclidean Spaces*, Princeton Univ. Press, Princeton, NJ, 1971.

82. G. Visick, *A weak majorization involving the matrices $A \circ B$ and AB*, Linear Algebra Appl., **223/224**(1995) 731-744.

83. B. L. van der Waerden, *Aufgabe 45*, Jahresber. d. D.M.V., **35**(1926) 117.

84. B. Y. Wang and F. Zhang, *Schur complements and matrix inequalities of Hadamard products*, Linear and Multilinear Algebra, **43**(1997) 315-326.

85. X. Zhan, *Inequalities for the singular values of Hadamard products*, SIAM J. Matrix Anal. Appl., **18**(1997) 1093-1095.

86. X. Zhan, *Inequalities involving Hadamard products and unitarily invariant norms*, Adv. Math. (China), **27**(1998) 416-422.

87. X. Zhan, *Inequalities for unitarily invariant norms*, SIAM J. Matrix Anal. Appl., **20**(1998) 466-470.

88. X. Zhan, *Norm inequalities for Cartesian decompositions*, Linear Algebra Appl., **286**(1999) 297-301.

89. X. Zhan, *Some research problems on the Hadamard product and singular values of matrices*, Linear and Multilinear Algebra, **47**(2000) 191-194.

90. X. Zhan, *Singular values of differences of positive semidefinite matrices*, SIAM J. Matrix Anal. Appl., **22**(2000) 819-823.

91. X. Zhan, *Linear preservers that permute the entries of a matrix*, Amer. Math. Monthly, **108**(2001) 643-645.

Index

Lecture Notes in Mathematics

For information about Vols. 1–1600
please contact your bookseller or Springer-Verlag

Vol. 1701: Ti-Jun Xiao, J. Liang, The Cauchy Problem of Higher Order Abstract Differential Equations, XII, 302 pages. 1998.

Vol. 1702: J. Ma, J. Yong, Forward-Backward Stochastic Differential Equations and Their Applications. XIII, 270 pages. 1999.

Vol. 1703: R. M. Dudley, R. Norvaiša, Differentiability of Six Operators on Nonsmooth Functions and p-Variation. VIII, 272 pages. 1999.

Vol. 1704: H. Tamanoi, Elliptic Genera and Vertex Operator Super-Algebras. VI, 390 pages. 1999.

Vol. 1705: I. Nikolaev, E. Zhuzhoma, Flows in 2-dimensional Manifolds. XIX, 294 pages. 1999.

Vol. 1706: S. Yu. Pilyugin, Shadowing in Dynamical Systems. XVII, 271 pages. 1999.

Vol. 1707: R. Pytlak, Numerical Methods for Optimal Control Problems with State Constraints. XV, 215 pages. 1999.

Vol. 1708: K. Zuo, Representations of Fundamental Groups of Algebraic Varieties. VII, 139 pages. 1999.

Vol. 1709: J. Azéma, M. Émery, M. Ledoux, M. Yor (Eds), Séminaire de Probabilités XXXIII. VIII, 418 pages. 1999.

Vol. 1710: M. Koecher, The Minnesota Notes on Jordan Algebras and Their Applications. IX, 173 pages. 1999.

Vol. 1711: W. Ricker, Operator Algebras Generated by Commuting Projections: A Vector Measure Approach. XVII, 159 pages. 1999.

Vol. 1712: N. Schwartz, J. J. Madden, Semi-algebraic Function Rings and Reflectors of Partially Ordered Rings. XI, 279 pages. 1999.

Vol. 1713: F. Bethuel, G. Huisken, S. Müller, K. Steffen, Calculus of Variations and Geometric Evolution Problems. Cetraro, 1996. Editors: S. Hildebrandt, M. Struwe. VII, 293 pages. 1999.

Vol. 1714: O. Diekmann, R. Durrett, K. P. Hadeler, P. K. Maini, H. L. Smith, Mathematics Inspired by Biology. Martina Franca, 1997. Editors: V. Capasso, O. Diekmann. VII, 268 pages. 1999.

Vol. 1715: N. V. Krylov, M. Röckner, J. Zabczyk, Stochastic PDE's and Kolmogorov Equations in Infinite Dimensions. Cetraro, 1998. Editor: G. Da Prato. VIII, 239 pages. 1999.

Vol. 1716: J. Coates, R. Greenberg, K. A. Ribet, K. Rubin, Arithmetic Theory of Elliptic Curves. Cetraro, 1997. Editor: C. Viola. VIII, 260 pages. 1999.

Vol. 1717: J. Bertoin, F. Martinelli, Y. Peres, Lectures on Probability Theory and Statistics. Saint-Flour, 1997. Editor: P. Bernard. IX, 291 pages. 1999.

Vol. 1718: A. Eberle, Uniqueness and Non-Uniqueness of Semigroups Generated by Singular Diffusion Operators. VIII, 262 pages. 1999.

Vol. 1719: K. R. Meyer, Periodic Solutions of the N-Body Problem. IX, 144 pages. 1999.

Vol. 1720: D. Elworthy, Y. Le Jan, X-M. Li, On the Geometry of Diffusion Operators and Stochastic Flows. IV, 118 pages. 1999.

Vol. 1721: A. Iarrobino, V. Kanev, Power Sums, Gorenstein Algebras, and Determinantal Loci. XXVII, 345 pages. 1999.

Vol. 1722: R. McCutcheon, Elemental Methods in Ergodic Ramsey Theory. VI, 160 pages. 1999.

Vol. 1723: J. P. Croisille, C. Lebeau, Diffraction by an Immersed Elastic Wedge. VI, 134 pages. 1999.

Vol. 1724: V. N. Kolokoltsov, Semiclassical Analysis for Diffusions and Stochastic Processes. VIII, 347 pages. 2000.

Vol. 1725: D. A. Wolf-Gladrow, Lattice-Gas Cellular Automata and Lattice Boltzmann Models. IX, 308 pages. 2000.

Vol. 1726: V. Marić, Regular Variation and Differential Equations. X, 127 pages. 2000.

Vol. 1727: P. Kravanja M. Van Barel, Computing the Zeros of Analytic Functions. VII, 111 pages. 2000.

Vol. 1728: K. Gatermann Computer Algebra Methods for Equivariant Dynamical Systems. XV, 153 pages. 2000.

Vol. 1729: J. Azéma, M. Émery, M. Ledoux, M. Yor Séminaire de Probabilités XXXIV. VI, 431 pages. 2000.

Vol. 1730: S. Graf, H. Luschgy, Foundations of Quantization for Probability Distributions. X, 230 pages. 2000.

Vol. 1731: T. Hsu, Quilts: Central Extensions, Braid Actions, and Finite Groups. XII, 185 pages. 2000.

Vol. 1732: K. Keller, Invariant Factors, Julia Equivalences and the (Abstract) Mandelbrot Set. X, 206 pages. 2000.

Vol. 1733: K. Ritter, Average-Case Analysis of Numerical Problems. IX, 254 pages. 2000.

Vol. 1734: M. Espedal, A. Fasano, A. Mikelić, Filtration in Porous Media and Industrial Applications. Cetraro 1998. Editor: A. Fasano. 2000.

Vol. 1735: D. Yafaev, Scattering Theory: Some Old and New Problems. XVI, 169 pages. 2000.

Vol. 1736: B. O. Turesson, Nonlinear Potential Theory and Weighted Sobolev Spaces. XIV, 173 pages. 2000.

Vol. 1737: S. Wakabayashi, Classical Microlocal Analysis in the Space of Hyperfunctions. VIII, 367 pages. 2000.

Vol. 1738: M. Émery, A. Nemirovski, D. Voiculescu, Lectures on Probability Theory and Statistics. XI, 356 pages. 2000.

Vol. 1739: R. Burkard, P. Deuflhard, A. Jameson, J.-L. Lions, G. Strang, Computational Mathematics Driven by Industrial Problems. Martina Franca, 1999. Editors: V. Capasso, H. Engl, J. Periaux. VII, 418 pages. 2000.

Vol. 1740: B. Kawohl, O. Pironneau, L. Tartar, J.-P. Zolesio, Optimal Shape Design. Tróia, Portugal 1999. Editors: A. Cellina, A. Ornelas. IX, 388 pages. 2000.

Vol. 1741: E. Lombardi, Oscillatory Integrals and Phenomena Beyond all Algebraic Orders. XV, 413 pages. 2000.

Vol. 1742: A. Unterberger, Quantization and Non-holomorphic Modular Forms. VIII, 253 pages. 2000.

Vol. 1743: L. Habermann, Riemannian Metrics of Constant Mass and Moduli Spaces of Conformal Structures. XII, 116 pages. 2000.

Vol. 1744: M. Kunze, Non-Smooth Dynamical Systems. X, 228 pages. 2000.

Vol. 1745: V. D. Milman, G. Schechtman, Geometric Aspects of Functional Analysis. VIII, 289 pages. 2000.

Vol. 1746: A. Degtyarev, I. Itenberg, V. Kharlamov, Real Enriques Surfaces. XVI, 259 pages. 2000.

Vol. 1747: L. W. Christensen, Gorenstein Dimensions. VIII, 204 pages. 2000.

Vol. 1748: M. Ruzicka, Electrorheological Fluids: Modeling and Mathematical Theory. XV, 176 pages. 2001.

Vol. 1749: M. Fuchs, G. Seregin, Variational Methods for Problems from Plasticity Theory and for Generalized Newtonian Fluids. VI, 269 pages. 2001.

Vol. 1750: B. Conrad, Grothendieck Duality and Base Change. X, 296 pages. 2001.

Vol. 1751: N. J. Cutland, Loeb Measures in Practice: Recent Advances. XI, 111 pages. 2001.

Recent Reprints and New Editions